The Energy of Nations

To understand what is going on you need to be a polymath who has worked at the highest levels on all sides. Jeremy Leggett is that person, and he provides clarity of thinking in a consistently delightful written style.

Paul Dickinson, Chairman, Carbon Disclosure Project

It really is a terrific read – very unveiling of our human struggle between greed and legacy.

Stephan Dolezalek, Managing Director, Vantage Point Capital Partners

I was so captivated by it. The narrative device of running the years and the oil price gives it power and tension. The inter-weaving of account and candid diary makes one feel one has a front-row seat at the places where things almost happen.

Adam Poole, Analyst, Buro-Happold

I was hooked within the first few pages.

Pamela Hartigan, Director, Skoll Centre for Social Entrepreneurship, Said Business School, University of Oxford

At a time when most people view tomorrow's energy prospects through dark lenses of coal, oil, fracked gas or even methane hydrates, Jeremy Leggett shines a brilliant light on the path towards low or zero carbon energy. Illuminating. And a joy to read.

John Elkington, cofounder of ENDS, SustainAbility and Volans

Brilliant – a real roadmap to the future. And a perfect reminder of why we must leave most fossil fuel safely in the ground.

Bill McKibben, Founder, 350.org

Jeremy Leggett is one of great entrepreneurs of the emerging solar era, a man driven by his passion for the environment and for social justice in the developing world to set up a new business (and a new charity) to give expression to those ideas. *The Energy of Nations* tells it as it needs to be heard, with new – and genuinely sustainable – business models at the heart of the transformation going on today in the global economy.

Jonathon Porritt, Founder Director, Forum for the Future

The Energy of Nations

Risk blindness and the
road to renaissance

Jeremy Leggett

Routledge
Taylor & Francis Group

LONDON AND NEW YORK

from Routledge

First edition published 2014
by Routledge

Simultaneously published in the USA and Canada
by Routledge
711 Third Avenue, New York, NY 10017

Routledge is an imprint of the Taylor & Francis Group, an informa business

British Library Cataloguing in Publication Data
A catalogue record for this book is available from the British Library

Library of Congress Cataloging-in-Publication Data
Leggett, Jeremy K.
    The energy of nations : risk blindness and the road to renaissance /
Jeremy Leggett. – First edition.
        pages cm
    Includes bibliographical references and index.
    1. Renewable energy resources–Social aspects. 2. Energy
development–Social aspects. 3. Energy consumption–Social aspects.
4. Petroleum reserves. I. Title.
    TJ808.L44 2014
    333.79'4–dc23   2013016393

ISBN13: 978–0–415–85782–6 (pbk)
ISBN13: 978–1–315–88741–8 (ebk)

Typeset in Garamond
by Keystroke, Station Road, Codsall, Wolverhampton

Printed and bound in Great Britain by
TJ International Ltd, Padstow, Cornwall

Contents

About the author

Jeremy Leggett is a social entrepreneur and author of *The Carbon War*, *Half Gone* and *The Solar Century*. Described by the *Observer* as 'Britain's most respected green energy boss', he has been a CNN Principal Voice, and an Entrepreneur of the Year at the New Energy Awards. He is founder and chairman of Solarcentury, the UK's fastest growing renewable energy company since 2000, and founder and chairman of SolarAid, an African solar lighting charity set up with 5% of Solarcentury's annual profits and itself parent to a social venture, SunnyMoney, that is the top-selling retailer of solar lights in Africa. He was the first Hillary Laureate for International Leadership on Climate Change, chairs the financial-sector think-tank Carbon Tracker and is a consultant on systemic risk to large corporations. He writes and blogs on occasion for the *Guardian* and the *Financial Times*, lectures on short courses in business and society at the universities of Cambridge and St Gallen, and is an Associate Fellow at Oxford University's Environmental Change Institute.

Note about the author's credentials and motivations

In this book Jeremy Leggett warns of five global systemic risks in or connected with the energy sector, each of which he has deep experience of. The first involves the risk of an oil shock. He was a creature of the oil industry for the decade of the 1980s, teaching petroleum geologists and petroleum engineers in the Royal School of Mines at Imperial College, consulting widely in the industry around the world, and researching geological history, including work on shale – oil and gas source rock – funded by BP and Shell among others. In 2006, his concerns about the potential for a mismatch between global oil demand and supply led him to convene the UK Industry Taskforce on Peak Oil and Energy Security.

The second risk Jeremy writes about involves climate shock: the potential for ruinous economic and environmental impacts of a rising global thermostat, driven by soaring greenhouse-gas emissions, mostly emanating from the energy sector. His research at Imperial College involved the geological history of the oceans, and hence ancient climates. He won major international awards for that work and was appointed a Reader at the remarkably young age of 33. His discoveries drove him to a concern about global warming so great that he quit academia and industry in 1989 to become an environmental campaigner.

He won the US Climate Institute's Award for Advancing Understanding for his work in the 1990s, campaigning for action on climate change.

In 1997, his continuing concern motivated him to set up a solar energy company to campaign directly in the markets. Solarcentury's purpose is to make as big a difference as possible in combating climate change. That purpose, around which the company's famously strong culture is built, was first pitched to investors in its initial capitalisation, in 2000. After four rounds of venture capital, the last in 2007, the company's purpose remains the same.

The third risk he writes about involves a further crash in the global financial system. One might think a social entrepreneur would have little to say here that would be rooted in experience. In fact, Jeremy has been a non-executive director of a Swiss bank's private equity fund, investing in renewable energy, since 2000. He has witnessed the financial sector at work from inside a bank's boardroom in the run up to, during and after the credit crunch and financial crash. He knows exactly how ruinous that episode was for many companies. His fund, New Energies Invest AG, was faring relatively well until 2007. As credit dried up, one by one the companies in the portfolio went under. New Energies Invest is now in liquidation.

The fourth risk he writes of is a carbon bubble in the capital markets. Companies and stock exchanges are currently being allowed to account coal, oil and gas reserves as assets at zero risk of stranding by climate policymaking. As a result, such is the enormity of carbon-fuel-based value on stock exchanges that the risk of systemic financial failure builds with every new reserve of fossil fuel discovered. Jeremy chairs Carbon Tracker, a small think-tank of analysts from within and around the City of London, experts who have worked for institutions like Henderson Global Investors and PWC. Some foundation funders hold the view that Carbon Tracker's whistleblowing on the carbon bubble may be the single most promising route to progress in slowing the climatically ruinous flow of capital to fossil fuels.

The fifth risk he describes entails the shale gas boom, now extending into oil from shale, what the industry calls tight oil. Notwithstanding all the PR from the oil industry about shale gas and tight oil being game-changers, great for the US economy and good for the world, this too is at risk of being another giant bubble waiting to burst, in Jeremy's view. As an award-winning research geologist, one of Jeremy's fields of expertise involved shale. He was a graduate student and faculty member at the same time as some of the current captains of the global oil industry, themselves also geologists. At the *Financial Times* Global Energy Summit in 2012, he debated with two enthusiastic shale gas and tight oil proponents: Exxon's European boss, and former BP CEO Tony Hayward, whom he has known since his youth. The video of that tense debate can be seen on the *FT*'s website.

We commend this book. This author both knows what he is talking about and is putting his money – not to mention his life's work – where his mouth is.

Prologue

No sooner had the twenty-first century opened for play than the global economy suffered its first crash. Most of us watching the dot.com bubble burst in January 2000, and then sweating through the recession that followed it, probably thought it would be a long time before the players in the capital markets made such a mistake again. With the benefit of hindsight, the post-mortems on the crash seemed so clear. Sensible people had been investing vast sums in paper ideas for making money via internet sales and advertising, years after burning many millions in advertising-led scrambles to position their brands ahead of rivals. As the bubble in internet stocks inflated, many investors or advisers of investors had been gaming the system – buying stock in companies producing nothing, valued way higher than long-lived pillars of the stock exchanges who were delivering real goods and services, hoping to sell on to greater fools before the inevitable crash came. 'Such a piece of crap', one investment banker wrote of a stock in a famous e-mail. It didn't stop him advising clients to buy it.[1]

When that kind of behaviour reached its natural endgame, giant corporations plunged in value by two thousand billion dollars overnight.

By 2006 the new craze involved financial products. Some of us may have tried to understand exactly what collateralised debt obligations and other such derivatives were. The vast majority, myself included, were prepared to believe the new doctrine. It held that clever investment bankers had found a way to limit or even excise risk from securities backed by 'sub-prime' mortgages: those extended to less well-off people. To the extent that many who otherwise might never have owned homes could now buy them, this was even a good thing. So the comforting narrative ran.

Some had an alternative narrative, when they could make their voices heard, and it was very uncomfortable. Dissident economists and financial journalists worried that the derivatives were actually compounding risk in the markets, that borrowing the vast amounts of money needed to create sub-prime mortgages was unsustainable, and that securities based on these mortgages would ultimately prove toxic as ever-growing numbers of people found themselves unable to service their debts. But these people were few, and the incumbency was so very quick to pour scorn on them. Scaremongers, they were deemed. Even if they were distinguished economists, like Nouriel Roubini, or worked for the *Financial Times*, like Gillian Tett.

By August 2007, watching credit markets freeze in a wildfire outbreak of panic that another bubble had been created, we knew the scaremongers had if anything been insufficiently strident in their whistleblowing. By August 2008, as the financial markets crashed, we realised that the incumbency had assembled a time bomb primed for economic calamity.

Fortunately, world leaders proved capable, under the gun, of heading off the immediate worst of possibilities: complete meltdown of the global financial system.

Now the post-mortems began again. Once more, the mistakes seemed obvious in hindsight. CEOs of investment banks raised their hands in Congress to swear that they too had not understood what collateralised debt obligations were. They were just the CEOs, we were to understand, not the rocket scientists who had designed the 'products'. Ratings agency staff had been too busy going on ski holidays with bankers, and worse, to do anything other than give these products triple-A ratings. And so on. Library shelves have already been filled with books on the lessons learned.

Political leaders and regulators have had four years since the great financial crash of 2008 to sift out further systemic risks to capital markets. How have they been faring?

Not well. At the time of writing in April 2013, systemic financial risk marches on almost unreconstructed in the financial services industry, despite everything we have learned about our collective ability to believe comforting narratives and ignore uncomfortable alternative views.

What about other sectors of the global economy?

Energy is the sector I work in. Since the dot.com crash, I have developed an ex-academic's passion for studying the patterns of play in energy markets, and in the financial markets where they pertain to energy. I have logged and analysed these patterns in a 2005 book, *Half Gone,* and since 2006 on my website www.jeremyleggett.net. As a result, today I count four global systemic risks directly connected to energy that threaten capital markets and hence the global economy. They involve oil depletion, carbon emissions, carbon assets, and shale gas. A market shock involving any one these would be capable of triggering a tsunami of economic and social problems, and, of course, there is no law of economics that says only one can hit at one time. There are other systemic risks in the energy sector, of course. Persistence with and proliferation of nuclear power risk spawning nuclear devices with which terrorists could take out cities, or weapons that could make nuclear war between nations more likely. The growing use of water in most forms of energy production could accelerate an already grave global water crisis. And so on. But in this book I want to concentrate on the four systemic risks that I have most direct vocational experience of.

I also write about a fifth risk. Ongoing systemic risk in the financial sector may at first glance seem to have nothing to do with energy. But I share a common view, fashioned with the benefit of hindsight, as are the views of so many interested in the 2007 credit crunch and the 2008 financial crash, that these were indirectly connected to energy. Peak panic about the toxicity of mortgage-backed securities followed shortly after the highest ever oil price in history: $147 per barrel in July 2008.[2] Could the high price of gasoline have had anything to do with all those owners of shiny new homes in American suburbs defaulting on their sub-prime mortgages? How could it not?

How best to describe the risks, and analyse the dangers? I have chosen to do this in the course of a historical account, rooted in my own experiences. I hope a chronological narrative approach will interest the reader more than a conventional format, while giving me the opportunity to recount how my own thinking has evolved over the years in the stoking of the serial crisis that faces society today. I tell that story in Part I: A History. Along the way, and especially in the last chapter, I discuss why it is that so many people choose to go on believing comforting narratives, ignoring or downplaying uncomfortable alternative narratives like the five I warn of.

In Part II: A Future, I set out my best-guess scenario for how this five-pronged set of dramas will play out in the years ahead, and analyse the implications. Amid the mega-risks, there are also opportunities for society. As the story unfolds, a potential good-news future scenario emerges that I think of as 'renaissance'. In Part II, I explain what I mean by that, and how I think it can be achieved. In the closing chapter, I assess our chances of achieving it, or some version of it, as opposed to the very bleak future scenarios that we risk by maintaining our present course.

Before starting my historical account, let me summarise each of the five risk issues quickly, aiming to provide the reader with a clear statement of the core contentions ahead of the narrative. I am hoping the unfolding of the story will then allow the reader a defensible judgement about the probabilities involved in each case.

The first risk I consider is a crash resulting from oil depletion. A minority of people in and around the oil industry worry that global supply of oil will cease growing within just a few years – reach a peak of production, in other words – and then start to fall, becoming unable to meet global demand. If we are correct, and nothing is done to soften the landing, the twenty-first century is almost certainly heading for an early depression. But the dominant view, held in governments, boardrooms and households alike – sometimes explicitly but mostly by default – is that peak oil is far off: that oil supply can continue to grow, fuelling the demands of a growing global economy, if necessary for decades to come.

The US shale gas boom, whereby huge volumes of gas have been produced by the process of fracking over the last decade, has done much to strengthen this comforting view. Fracking is now being applied to extract oil from shale, and is lifting domestic US production spectacularly, against many expectations. Some even hope to see 'Saudi America' emerge in the years ahead: a USA leading the world in production and even self-sufficient in hydrocarbons. The incumbency – by which I mean most of the oil and gas industries, their financiers, and their supporters and defenders in public service – do everything they can to advance this comforting narrative. They also profess that the phenomenon will be exported, in time, to the rest of the world. This, they tend to say, has confounded the 'peakists' and is even the death of peak oil as an idea.

The second risk involves a further financial shock. Growing numbers of financial experts are warning that failure to rein in the financial sector in the aftermath of the financial crash of 2008 makes a second crash almost inevitable. Others argue, however, that modern capitalism is cyclical and resilient, and that recovery can be expected if the regulatory touch remains light and austerity budgets are applied in a concerted way by governments.

The third risk involves a crash related to climate change. For many observers, climate change, driven by the radiative forcing of greenhouse-gas emissions, is progressing faster than even the gloomiest forecasts of climate scientists a few years ago, and emerging impacts of a destabilising climate foreshadow major future economic disaster potential, for example in global food and water supply. For others, climate changes are to be expected, are not driven by mankind's emissions, adaptation is possible, and there is little or no need for concern.

The fourth risk involves a carbon asset bubble in the capital markets. When analysts convert reserves of fossil fuels into carbon dioxide emission-equivalent, and compare the total to a global carbon budget for keeping global warming below a widely acknowledged danger threshold of a 2°C rise in the global average temperature, a disturbing set of figures emerges. There is far more carbon in fossil fuels, in this way of looking at things, than society can afford to burn. Despite this overshoot of 'unburnable carbon' building on the capital markets, carbon-fuel companies go on turning resources into reserves, and investors continue to pile in. If governments do at some stage begin to panic about the emerging impacts of climate change, and summon the collective regulatory will to do something about cutting emissions, a lot of supposed assets are going to suffer an immediate erosion of value. And that is the stuff of potential financial shock.

The fossil-fuel incumbency relies on there continuing to be no restrictions. The massive part of the financial sector that is an enthusiastic player in this incumbency is effectively saying to governments that 'we do not believe you will ever do anything to limit carbon emissions – not even a fraction of what you say you will do'. And so the bubble inflates and the danger builds.

The fifth risk involves a shale gas crisis. For many, the explosive growth of shale gas production in the USA – now extending into oil from shale, or 'tight oil' as it is properly known – is a revolution, a game-changer, and it even heralds a 'new era of fossil fuels'. For a minority, it shows all the signs of being the next bubble in the markets. In the incumbency's widely held view, the US shale gas phenomenon can be exported, opening the way to cheap gas in multiple countries. For others, even if there is no bubble, the phenomenon is not particularly exportable, for a range of environmental, economic and political reasons.

This risk too entails shock potential. Take a country like the UK. Its Treasury wishes actively to suppress renewables, so as to ensure that investors won't be deterred from bankrolling the conversion of the UK into a 'gas hub'. Picture the scene if most of the national energy eggs are put in that basket, infrastructure is capitalised, and then supplies of cheap gas fall far short of requirement, or even fail to materialise.

How do the odds stack up in these five interconnected systemic global risk debates? How might they interplay in the unfolding events of the 2010s?

This book offers one person's view.

Acknowledgements

Many people read sections of this book to help me iron out glitches. You are too many to mention, but you know who you are, and you know you have my sincere appreciation. Those who read the whole book and gave me feedback include Roger Bentley, Peter Colville, Stephan Dolezalek, Colin Hines, Mark Lewis, Nick Robins and Stephan Schmidheiny. I am especially grateful to them. However, hardly any two people agree on everything when it comes to energy and its wider systemic context, so it should not be assumed that they agree with me on every point. And of course, nobody but me has responsibility for any mistakes that might have slipped through.

Note on sources, style and the life of the project beyond the book

I reference all direct quotes in the endnotes. All other sources can easily be located on my website, www.jeremyleggett.net, using the word-search facility.

Aiming as I am for as wide a non-expert readership as possible, I have tried not to write a technical book. There are plenty of those in the energy field, and I reference some key ones as I go. As an ex-academic I was often tempted to go into further detail to help make my point, especially on the detail of oil supply. But as a writer I knew that would quickly make the book dense and easy to put down. In these places I use endnotes to refer the reader to entries on my website that go into more detail, or I give links to vital work by others that expands on the relevant technical arguments.

The diary extracts recount real events and conversations, but I made no tape recordings at them, so the dialogue is from memory, usually written up in my diary immediately after the scenes described. I vouch completely for the sense of the dialogue, but obviously cannot vouch for the exact words. Hence I use no quotation marks. I also use extracts from conversations, not the whole. When I do that, I quote nothing out of context.

Where I quote from events filmed on TV or YouTube and elsewhere in the media, I reference the link, so the reader can hear the whole.

Any book like this about fast-moving dramas is out of date as soon as it is written. But the dramas I chronicle and analyse in the book can be followed on my website, www.jeremyleggett.net, where I will be keeping them up to date. As for my analysis of events, I will keep writing about what I think the unfolding history means, both on my website and in a monthly column in *Recharge* magazine.

Part I

A history

Chapter 1

Lies, scaremongering and affordable oil

Imagine you are the CEO of Shell, and you receive an e-mail from your Head of Exploration complaining that he is sick and tired of lying about the company's oil reserves. Would you fear your days as CEO were numbered? Would the prospect of a prison cell intrude on your comfortable life?

In November 2003, the then CEO of Shell, Phil Watts, did receive such an e-mail from Shell's then exploration chief, Walter van de Vijver. 'I am becoming sick and tired of lying about the extent of our reserves issues and the downward revisions that need to be done because of far too aggressive/optimistic bookings', it read.[1]

In April 2004, that e-mail went public. Watts had admitted in January that Shell's reserves were overstated by 20%. In the months to come, he would have to revise them down a further three times. He was fired. But that was the least of his troubles. The US Justice Department was now conducting a criminal investigation and the British Financial Services Authority was after him too.[2]

2004 was a bad year to be caught telling lies about oil reserves. The oil price had wandered along at around $20 for the best part of two decades, but it was now starting to climb, and fast. It hit $40 for the first time in May. Pain quickly spread around the world. American consumers found themselves paying an extra $44 billion at the pump during the first half of the year. They and their thinning wallets began staying out of the shopping malls, and so the misery spread to retailers. As one retail executive saw it: 'We are hurt by high oil prices because people are giving their extra dollars to Exxon.'[3]

Shell had created jitters in the market with its lies, but there were other reasons for the then record prices. The war in Iraq was entering its second year. It was not going well. With insurgents attacking oil pipelines on a seemingly daily basis, maintaining exports was proving impossible. The Abu Ghraib torture scandal broke in May, fanning the flames further. In June, terrorists attacked oil infrastructure within Saudi Arabia, the world's number one producer, the nation with the largest reserves by far, spreading the jitters to the main Saudi oil-export terminal at Ras Tanura. Said one analyst: 'If you can blow up the Pentagon in broad daylight, then it cannot be impossible to fly a plane into Ras Tanura – and then you are talking $100 [per barrel of] oil.'[4]

Looking back from the vantage point of 2013, with an average oil price in 2012 of $111, we can only wonder what the price would be driven to if Ras Tanura blew up today.

The year 2004 proved to be the beginning of a long wobbly climb in the oil price that, in the decade ahead, would change the world. There is a lot you can't do with $100 oil that you can do with $20 oil: like buy a suburban house on a low salary, for example. Equally, there is a lot you can do with $100 oil that you can't do with $20 oil: like grow renewable energy markets, for example.

What is supposed to happen in situations of tight global oil supply like this is that OPEC, the Organisation of the Petroleum Exporting Countries, opens up its spare capacity – its ability to pump more oil, held in reserve – floods the market and cools the oil price. In August 2004, OPEC announced that it no longer had any spare capacity left to tap. The price duly moved up to a new record near $45.

At that point headline writers began flagging a threat to the world economy.[5]

It seems incredible looking back. But such is the importance of affordable oil in our oil-overdependent global economy. A little panic about unafford-ability goes a long way.

The markets themselves weren't helping. Hedge funds had begun betting on a high oil price. When this happens, the potential for self-fulfilling prophecy comes into play.

Investors grilled BP, fearing it might have been playing the same game as Shell. BP's CEO Lord John Browne gave a series of speeches aimed at reassuring the financial world that BP was not Shell, and that both his company and the wider oil industry could be relied on to deliver growing supply from real reserves.

'There isn't really a supply crunch at the moment', he asserted. 'We have the perception of the risk of a supply interruption, but that's all we've got.'[6] Indeed, he said, there is 40 years of supply.

This is a statement that BP have repeated many times since. They derive the figure by dividing the total global reserves reported by oil companies, more than 1,000 billion barrels, by the global demand. The mantra sows comfort while neatly sidestepping the underlying concern.

All through 2004, those concerned about the rate of oil depletion sought to warn about the risk of peak oil. Our oil-dependent global economy is at risk of a crash, should it discover that oil supply cannot rise in line with demand. The persistently high oil price gave us a context, and hence the chance for a degree of unfamiliar exposure.

Oil reserves under the ground are not the same as oil flows from production pipes at the surface, we explained. Oil resources, deposits of oil extractable in theory, are not the same as oil reserves, deposits of oil mapped out and extractable economically. Peak oil is the point at which the depletion of existing oil reserves around the world can no longer be replaced by additions of new flow capacity. Oil production reaches the highest level it ever will, and drops. It can drop for what we can think of as below-ground reasons or for above-ground reasons, or both. Below-ground reasons involve the geology of depletion: how much oil there really is down there, and how fast we suck it out. Above-ground reasons involve geopolitics, the behaviour of nations and their citizens, which can so easily mess up oil production. Most of us think that both below- and above-ground factors will define the peak, but the peak will be the peak: the most oil that can ever be produced in any one day. Perhaps production will wobble along on a plateau for a while before dropping, but it will never exceed that peak level. If we think of all the theoretically extractable oil under the ground as a tank, what we have to worry about is not so much the size of that tank, but the size of the taps: the actual global oil production capacity.

The oil industry usually doesn't question the fact that some day there will be a peak in production: how can it, when oil deposits are finite? It simply says the peak will happen much later. In its over-exuberant rhetoric, the industry constantly focuses attention on its estimates of the size of the tank: both the 40 years of supply in the reserves, and the vastly higher figure in the resources, those deposits that they have yet to prove they can extract economically. Those who fear an early oil-production peak worry about the flow rates the industry can deliver from the actual taps in place today. If the taps start running slower, and the industry can't meet global oil demand, we have a crisis on our hands. We live in a system where markets are prone to panic. We live in a society where food supply needs lots of affordable oil all the way from the field to the plate.

There is also, as it happens, reason to be worried about the size of the tank. OPEC governments, including Saudi Arabia, have been less than transparent about the size of their national reserves since deciding to fix quotas based on the size of those reserves in the 1980s. Some experts, including within OPEC itself, profess that at least 300 billion barrels out of the 1.2 trillion barrels of supposed global proved reserves may have been overstated.

Sceptical oilmen refer to these barrels as 'political oil'. Others use less polite language to describe the situation.

Why don't people check, one might ask?

And there is the dilemma. A dozen good geologists *could* check, and set this nasty suspicion to rest – or not as the case may be – in a matter of months. But the OPEC governments do not allow them into their oilfields to do so. That was the case in 2004. It remains so today.

In September 2004, the CEO of Total, Thierry Desmarest, called on the Saudis to let the oil majors in to help them boost output, so as to reduce the oil price. His offer went unheeded.

A revealing letter written by the First Secretary for Energy and Environment in the British Embassy in Washington found its way to me and others worried about peak oil at this time. The diplomat in question had attended a presentation on oil supply by the respected consultancy PFC. 'The presentation drew some gasps from the assembled energy cogniscenti', he reported back to London. 'They predict a peaking of global supply in the face of high demand by as early as 2015. This will lead to a more regional-ized oil market, a key role for West African producers, and continued high and volatile prices.'

Let me emphasise this year, at this early point in my account of events: 2015. We will find that it recurs in the story to come.

In October 2004, G7 finance ministers met in Washington for the annual meeting of the World Bank and the International Monetary Fund (IMF). They seemed as worried about oil supply as the diplomat. The closing statement by ministers and central bank governors read: 'Oil prices remain high and are a risk. So first, we call on oil producers to provide adequate supplies to ensure that prices remain moderate.'

As a veteran financial correspondent described it, 'my sense of [the] meetings is that there is an atmosphere of suppressed panic about the oil price, and about the danger of a serious crisis.'[7]

By the end of the month the oil price had crossed $55.

OPEC now called on the US to open its 670-million-barrel Strategic Petroleum Reserve to help suppress the prices. This reserve, stored in caverns in Louisiana salt mines, is meant to be on hold for major emergencies, not the cooling of price rises.

How was that for a signal that all may not be well in the Middle East, we early peak worriers asked. If Saudi Arabia doesn't have the spare capacity to cool a price rise, how can we be sure of its reported reserves?

BP CEO Lord Browne upped his rhetoric to try to calm the jitters. 'It is not helpful for the world to believe that it is running out of oil', he said. 'We are evidently not.'[8]

The early peakers read this in exasperation. This risk debate is not about the oil 'running out'. It will never run out. Oil reserves under the ground, we tried to say, once again, are not the same as oil flows from production pipes at the surface.

Somehow, we didn't seem to be able to get our messages heard as readily as BP did theirs. Even in a world where Shell had been caught lying about its reserves.

St James Park, London, October 2004

Tony Blair's man is in relaxed mood as we walk by the lake in the park, a short walk from his office in Number Ten on a lovely autumn morning. He has a smile on his face. I am more used to frowns from people such as he.

The visit worked really well, Jeremy, he says. We are all pleased.

And no doubt you are particularly pleased with yourself, I think but don't say, given that the whole thing was your idea.

The idea was this. The Prime Minister intends to top the agenda with climate change when he chairs next year's G8 Summit, in Scotland. On the day he announced this, he wanted an appropriately green and youthful photo-opportunity. His man suggested my company, just ten minutes away.

Blair is trying to persuade America to join Europe in efforts to cut greenhouse-gas emissions. He has sanctioned his Chief Scientific Adviser, Sir David King, to fire a test missile aimed at US opposition to cuts. Climate change is the most serious problem we are facing today, King has said: more serious even than the threat of terrorism.

That would make it more important than the war in Iraq then, I thought when I heard this. Because that is what the war on terror is about, supposedly.

On his visit to Solarcentury, Blair led a round-table discussion with a handful of the younger staff, television cameras and radio microphones trained on them for the first time in their lives.

What shall we ask him? they demanded of me as the Anti-Terrorist Branch's sniffer dogs descended on the office ahead of the visit, scrabbling around under their desks with vibrating noses.

Anything you like, I said.

Then, succumbing to a corporate afterthought: OK, it's up to you, but actually, perhaps, er, today wouldn't be a great day to mention the war.

These days, in my chosen path, I am discovering what many an entrepreneur does. It is difficult indeed to wear your principles on your sleeve.

After two hours in the office, I walked out onto the street with Blair, both of us expecting to see his motorcade there. But Lower Marsh is a busy market street. The cars were parked a hundred yards up the road. No security men were waiting. None had come out of the office with us. To my amazement, I walked alone along a crowded market with the Prime Minister as he nodded at flabbergasted people, his trademark beam full on.

Blair's man and I have lunch, talking climate politics and the prospects for success at the G8 summit. He tells me he feels a certain optimism. Blair and Bush are getting on well. There may be a chance of the Americans exercising some quid pro quo over the British support in Iraq. It's game on.

This, I think, is as good a cue as I will get.

Yes, Iraq, I say. Where oil production is shot to pieces, what with all the Shock-and-Awe. If the West loses Iraqi production and access to their reserves long term, what happens if the others in OPEC don't prove able to plug the gap? Do you worry about global oil production peaking?

The oil companies don't, says the PM's man. But I know that you and a few others do.

Even if it is only a few, I say, which it isn't, it's such a high-consequence risk, isn't it?

We bat a little detail around for a while. He knows a lot more than the basics, I find. He doesn't tell me he agrees with me that there is a problem. Neither does he say he disagrees.

I ask him how concerned citizens like me can best register our concerns about peak oil in mainstream political debate. The issue languishes, as he knows, many miles behind the appreciation climate change now has in the minds of publics and politicians.

He considers this a while, looking out of the window. I fancy that he is pondering how best to dissemble. He is a man who knows how to dissemble.

The problem you have, Jeremy, is that there is nothing in it for politicians.

I find his answer amazing. I ask him to explain.

If the early peak argument is right, and the peak and its shock hit while you're in office, you're dead. The opposition lie machine will pin the oil crash on you, and there will be nothing you can do to persuade the tabloid-reading public otherwise. On the other hand, if you believe in the early peak while in opposition, and try to warn about it, then you will be accused of irresponsible scaremongering, both by the energy industry and the sitting government. The tabloid press will crucify you. Oh, and the voters will hate you for telling them an unhappy story about the future.

So, I say, if I've got this right, you just sit tight and hope desperately that the likes of BP have got their story correct?

He shrugs, with that Latin use of the hands that says so hey, what can you do?

No dissembling today then.

I note that he seems to think it's all rather amusing.

I wonder if that is a requirement for survival in his world.

BP headquarters, St James's Square, London, 11 November 2004

Tony Hayward and I over croissants, coffee and fruit, immaculately served on the vast table in BP's boardroom. What dramas have unfolded in this room, I am thinking. What dramas will.

Hayward and I wore our hair long and curly when we first met as young geologists in the 1980s, researching the same kinds of rocks, from different universities. Now he heads all BP's upstream operations, two-thirds of the company, and is one of a handful of stars contending to succeed the current CEO.

I want to discuss my concerns about oil depletion with him. He is cool with that and lounges now in a fabulous suit, his boyish smile in place. His aura is one of confidence edged with shyness, just like it was a quarter of a century ago.

I wonder why he decided we should have breakfast in this cavernous room, not his office.

We joke about our past, exchanging news of old student friends. But my purpose today is far from a joking matter. During the year, the biggest scandal in British corporate history has unfolded. I fear that if one company is in trouble in the oil reserves department, why not others? Why not the oil industry as a whole? I worry that the Shell fiasco foreshadows future bad news about the global peak of oil production. BP's view is that there is nothing to worry about. I want to look my old student friend in the eye and hear why they believe that, exactly.

BP has of late been pouring money into Russia. But President Putin's government has recently launched a deadly legal assault on Russian oil giant Yukos. It is a thinly disguised re-nationalisation of domestic oil. I elect to begin with that little piece of energy insecurity.

Aren't President Putin's antics a danger to BP's prospects? I ask.

No, says Hayward firmly. We have a great relationship with the man himself, and those around him. I fly to Russia once a month or so, specifically to keep the relationships there warm.

I marvel at this. If I, as a solar industry boss, had a potentially mission-critical relationship with a single key player in foreign parts, and I told my board of directors not to worry because I was pals with him and saw him regularly, they would scoff in my face.

I drop the oblique approach and ask him straight.

Tony, to what if any extent do you worry that the minority arguing peak oil will come sooner rather than later might in fact be correct? After all, this is about risk, not certainty, huh?

The peak oilers are scaremongers.

He looks me in the eye as he uses the big 'S' word. He knows I am one of them. His smile doesn't shift.

But where is all that extra supply going to come from? I ask. The reserves in existing oilfields are depleting fast. The peak of new-oilfield discovery was way back in the mid-1960s. The industry isn't finding anything like as many giant fields as it did. The existing giant fields are mostly ancient. And something is sure to go wrong above ground too. Maybe already has in Iraq.

There is plenty of oil, he says emphatically, trust me on this.

I hear what you say Tony, and you more than anyone should know. But where?

The Middle East and Siberia.

I note that he doesn't hesitate in answering, or add Africa, the Americas or Asia to the list. Nor the tar sands in Canada.

When production peaks, he says, it will be because of global demand reduction, not the industry's inability to meet growing demand.

I wonder at the confidence he has in this too. He has been to China and India. He knows how fast those economies are growing, and the extent to which they have copied the West's oil dependency.

Questions jostle in a mental queue. I only have a small slot in Hayward's calendar. I'll never be able to ask them all, and I doubt I'll get another audience. I also want to see where his head is on global warming.

BP's oil production is a tiny fraction of the global total, I say. What happens if others let you down, for whatever reason, and the industry as a whole can't lift its production?

Then a huge recession would be unavoidable, Hayward says. But that just isn't going to happen.

A huge recession is one way to put it, I reflect. When one considers that the energy locked into a single barrel of oil is equivalent to the energy expended by five labourers working 12-hour days non-stop for a year, a recession is not the only problem that leaps onto the radar screen if supply is suddenly rationed.

Well, I guess we're going to find out, I say. One way or the other.

I move on, reluctantly. I tell him about my recent trip to Berlin, in a delegation of scientists and business leaders accompanying the Queen on a state visit, there to discuss climate change at her request with German counterparts. In those discussions, I heard government scientists express fears of horrible destabilisation of the climate system unless reductions of greenhouse-gas emissions from coal, oil and gas burning begin soon. Especially coal, of which there is so much more than oil and gas. Things are looking very serious.

On this threat, Hayward is not going to disagree with me. BP has long since elected to concede the reality of global warming and the climate change it causes.

Government needs to get on and govern, he says. They need to lead on this one.

This is what all the carbon-fuel bosses say, I object. And governments say the reverse – at least, those that are taking the threat seriously. If only the carbon-fuel companies would stop their infernal efforts to undermine our efforts at policymaking, I get told, if only they would act voluntarily in the face of the threat, maybe we could get somewhere. This blame transference is one of the main ways the greenhouse trap stays shut. How do we break the impasse?

Tony Hayward looks at his watch. I am clearly trying his patience now.

Maybe the corporate jet is waiting to ferry him to Moscow.

Chapter 2

Under the volcano

In March 2005, the oil price hit a new high of $57. The futures exchange Nymex, where oil is sold in fixed prices into the future, saw its first hundred dollar oil trade. News leaked that the International Energy Agency would soon be calling for an international emergency plan to cut oil demand. Thirty-one American national security leaders advised President Bush to cut US oil demand as a matter of urgency.

Reasons to be concerned about an early peak in global oil production continued to emerge. In April, an analysis by the Bank of Montreal concluded that Saudi Arabia's main oilfield, Ghawar, was nearing its peak of production. Of the kingdom's then nine million barrels a day of production, fully five came from Ghawar. Some 90% of Saudi oil came from just seven fields, all averaging 45–50 years old.

In June, the Houston investment banker Matt Simmons – a man who had made a fortune in the oil trade – published a book alleging that the entirety of Saudi reserves was in danger of collapse.[1] Soon thereafter a worrying assessment came from within the kingdom itself, from a man perfectly placed to know if there were problems. Sadad al-Husseini had until his recent retirement run all exploration and production at Saudi Aramco, the national Saudi oil company. Saudi Arabian reserves were in no danger of immediate collapse, he said, but the kingdom would struggle to reach peak production of any more than 12 million barrels a day, and certainly would never be able to double its oil production to 20 million barrels a day, as many optimists professed it would.

al-Husseini has continued to speak out broadly on the side of those worrying about early peak oil ever since. It is incredible to me that the incumbency can manage to marginalise such a man. But they do.

At the time of writing, Saudi production stands at around nine million barrels a day. The highest its annual average has ever reached is just above 11.5 million barrels a day, in 2012.[2]

In August 2005, Hurricane Katrina hit New Orleans, creating devastation that would change many an American mind about the climate change issue. It also disrupted oil production across the Gulf of Mexico and pushed the oil price above $70 for the first time.

In September, the International Monetary Fund warned that oil prices this high were threatening the global economy.

The International Energy Agency (IEA) offered little hope that prices could come down easily. In its 2005 annual *World Energy Outlook*, the agency concluded that collective oil production in non-OPEC countries would peak 'right after 2010'. This meant, obviously, that much would depend on OPEC thereafter. It meant, the IEA said, that a business-as-usual approach to supply and demand was 'not sustainable': demand would have to be reduced.

The IEA, an agency set up by the OECD countries in 1974 after the first global oil crisis, had long been critical of those worrying about early peak oil. Now it was issuing warnings of its own about oil depletion.

Rimini, Italy, 30 October 2005

Jim Schlesinger looks out at his audience, a grand old man savouring his turn at the podium. Assembled before him are oil ministers, OPEC officials, IEA officials, UN officials and an eclectic mix of the rest of the world of oil. I am fascinated to hear what he has to say. He has been director of the CIA and Secretary of Defense. He is steeped in oil. Its geoscience. Its geopolitics. Its secrets.

As Secretary of Defense during the first global oil shock in 1973, Schlesinger threatened that the USA would invade the Arabian peninsula if the Saudis didn't restart the pumps they had shut down in anger over the Yom Kippur war, so creating the crisis. He openly admitted it later: 'I was prepared to seize Abu Dhabi. Something small. But nothing big. Militarily we could have seized one of the Arab states. And the plan did indeed scare and anger them. No, it wasn't just bravado. It was clearly intended as a warning. I think the Arabs were quite worried about it after '73.'[3]

They were worried at the time. The Saudis had their oilfields rigged to blow if the US acted on its sabre rattling. Saudi Arabia in 1973 could have looked like Kuwait in 1991.

Today, in a sleepy sunlit Italian seaside town, Schlesinger is issuing no threats, just a warning.

The peak-oil problem and the human response to it – or rather, the lack of collective response – remind him of the rumbles under Vesuvius, he tells us, and the reaction of Pompeii's doomed residents to those warning signs. The peak or plateau in global oil production is coming, he says. We don't know exactly when, but the probability is sooner rather than later. When oil companies discover oilfields these days, they are finding prairie dogs, not elephants. Some 80% of global production comes today from elephants discovered before 1970, and these fields are being rapidly pumped towards exhaustion. Yet demand is soaring. Sooner or later the prairie dogs aren't going to be able to keep up.

Political systems do not deal easily with long-term threats, even if they have a probability of 100%, Schlesinger warns. Economic horror will descend on the world if we do not plan ahead. The time to start is now. We are asleep at the wheel, like the citizens of Pompeii and Herculaneum were, looking up at their volcano and thinking that its quiescent history would be its destiny. They ignored the rumbles, and ended up buried.

Most people, and all governments, are in denial, he says in closing. Every time someone says the peak is far off, there is an audible sigh of relief. He singles out Daniel Yergin, chairman of the influential oil industry consultancy Cambridge Energy Research Associates (CERA), one of the oil industry's favourite cheerleaders, as a prime example.

Yergin is in the audience. So too, for reasons that are lost on me, is the Hollywood actress Sharon Stone.

Schlesigner knows Yergin will follow him at the podium and have the last word.

Yergin duly obliges. I don't see why human genius can't meet the challenge of keeping production growing, he says. Indeed, his company has access to a proprietary database of oil reserves run by the IHS Group that would be enough to convert any doubter to optimism.

We have to trust him on that. Access to the database costs a million dollars or more.

One of the oil geologists who worked with that database in its early years, Colin Campbell, is at the Rimini summit. He has kept his own version updated since. His view is the polar opposite of Yergin's. He has become the lead architect of peak-oil whistleblowing in and around the oil industry. He is founder of the main umbrella organisation for those of us concerned about early peak oil, the Association for the Study of Peak Oil.

Campbell has drafted a protocol aiming to defuse the peak-oil crisis. It advocates the simple expedient of demand management at the same rate as global depletion. He has been told by the organisers of the summit that the protocol will be launched as an official document of the summit at the end of the event. Mikhail Gorbachev is to do the honours in a plenary speech. It will be known as the Rimini Protocol.

But Gorbachev doesn't show up. Campbell isn't allocated a slot to speak in the main plenary. Somewhere, somehow, amid all the machinations involved in persuading the oil industry's glitterati to turn up at the event, the Rimini Protocol has become sidelined.

Instead, Daniel Yergin gives the closing speech, reciting the comforting mantras of the incumbency. Urging the citizens to carry on ignoring the rumblings under the volcano.

In February 2005, the Kyoto Protocol came into force. In this treaty, negotiated in Japan in December 1997 by almost all the world's governments, developed nations pledged to reduce their collective greenhouse-gas emissions by 5.2%, by a target window of 2008–2012. The agreement was a mere first step on the road to the deep cuts in emissions ultimately needed to have any chance of avoiding grave danger from global warming. It had nonetheless taken seven years for the required majority of governments to ratify it.

Those seven years of stalling had seen a constant drip of worrying scientific assessments. The latest development was particularly concerning: a compilation of seven million water temperature measurements down to more than 2,000 feet in the oceans showing unequivocal warming right down to that depth.

On top of this came evidence of acidification of the oceans. When carbon dioxide mixes with seawater, it forms a weak acid. A large scientific study in July 2005 pointed to catastrophic impacts on ocean ecosystems from the steadily increasing quantities of carbon dioxide being emitted, primarily from fossil-fuel burning.

Opinion polls were showing huge concern about global warming at this time. Europe was in the grip of the worst drought on record, fuelling the concerns. Tony Blair was determined to make progress. But at the G8 Summit in Scotland, he found the Bush administration doggedly unwilling to budge on their rejection of commitments to emission reductions.

The bad news rolled on, immune to the hopes and prejudices of politicians. Measurements across Siberia's bogs showed that melting permafrost was adding extra methane to the atmosphere. This was an example of what scientists call positive feedback: a natural amplifier of warming. Feedbacks like this were causing some scientists to conclude that global warming might be hitting a tipping point: a point of rapid acceleration. The Amazon basin was suffering its worst ever drought, for which scientists were blaming the warmer-than-normal North Atlantic. Two major Antarctic glaciers were found to be discharging three times faster than they were ten years ago.

By the time of the annual climate summit, held that year in Canada, Tony Blair was not the only national leader desperate for progress on greenhouse-gas emissions.

Montreal, 7 December 2005

An icy city, with Christmas in the air and the climate circus in town. The name of the game at the Montreal climate summit is action beyond the Kyoto 'commitment period' of 2008–2012. The vast majority of the 190 governments present want further commitments, deeper emission reductions, and progress on all the mechanisms for achieving that.

The Bush administration, unbelievably, wants the whole process of negotiation killed.

The USA has not even ratified the Kyoto Protocol, but still it wants to see the treaty and its processes stall, preferably crash.

Bush does not have his citizenry lined up behind him, it seems. Mingling among the thousands of attendees at the summit is a large and diverse group from American civil society, including state and local government officials, faith groups, environment groups, trade unions and development groups. American cities are being particularly vocal. Seattle is playing a lead role in the creation of the World Mayors' Council on Climate Change, an alliance of mayors from major cities around the world taking unilateral action to cut emissions.

The business community is here in numbers too, and not all of them are on the same page as the 'carbon club' of oil and coal companies. A growing number of corporations are urging action on climate change, and acting themselves. The evidence is becoming just too compelling for them. Among the recent actors is none other than WalMart, *bête noire* of many environmentalists. The trauma of Hurricane Katrina has converted its CEO, Lee Scott, to the view that emissions reductions are essential. If government won't lead, Scott says, then dammit, he will. After all, the corporation he leads is bigger than most governments.

He has committed his company to a massive programme of emissions reductions, aiming ultimately at zero carbon use in energy. He is pressuring the many companies in his supply chain to do likewise.

One of WalMart's subsidiaries is Gazeley Properties. Their CEO, John Duggan, is feeling the heat from WalMart headquarters. He wants to see the reason for all the fuss. He has asked if he can accompany me to the annual climate summit. I am happy to take him along, provided the poor guy doesn't mind hearing my views on the peak-oil risk to WalMart's supply chain as well. Gazeley builds warehouses for retail clients. Few industries are more dependent on affordable oil.

Duggan and I arrive at the convention centre where the governments are negotiating. As we queue at the security checks, a host of campaigners, mostly young, hand out pamphlets. Some are dressed as polar bears, others as Pacific islanders.

One handout is adorned with oil-like ink splashes. 'It's Time To Exxpose Exxon', the title reads. 'ExxonMobil funds junk science to deny the existence of global warming rather than take concrete steps to combat it.' The reader is exhorted to drive past Exxon and Mobil gas stations, not to buy Exxon stock, not to work for Exxon, and so on. There is a global effort under way to target this corporation, deemed by environmentalists the biggest of the 'carbon criminals'.

I wonder what John Duggan will make of all this. It is a world far removed from the one he is used to.

He asks me to show him Exxon and its allies at work inside the process.

He won't have to look far for allies. Harlan Watson was once a coal industry lobbyist, a key player in the carbon club, constantly working along-side Big Oil's lobbyists. He is now head of the US government delegation, the man who represents Bush himself.[4]

We make our way to a gathering of industry representatives. A statement is being prepared for presentation to ministers on behalf of the business community. At the table sit the representatives of dozens of trade organ-isations and corporations, and in the thick of them is Brian Flannery, Exxon's long-standing lead lobbyist.

As I have seen often before, Flannery is pushing his mission, along with all his fellow wreckers from American oil and coal, from within the International Chamber of Commerce. This is an organisation that has never needed any persuasion to let Big Oil speak as though it represents all species of business.

Duggan and I listen to the contributions from around the table. Whispering, I explain to him that in my days attending the climate negotiations, between 1990 and 1997, virtually all the business attendees were there to try to stall or wreck the process. Today, with the scientific evidence becoming ever clearer, things are very different.

Flannery argues for toothless wording in the Business Statement: an end product that won't embarrass the Bush administration. A representative of the Business Council for Sustainable Energy suggests that if Exxon gets its way, there should be an alternative statement to plenary on behalf of those who think differently. British energy utilities are members of this group. John Duggan sees clearly that the business world is irrevocably split on climate change.

I wonder if the firming science has had any impact on Flannery's own conscience. I have seen this man mistreat scientific evidence for his masters many times over the years. I have chronicled his antics, and those of the other lead oil and coal lobbyists. I have accused the carbon club, and their masters, of misrepresentation, manipulation and distortion at best, and lying at worst.[5]

I approach Flannery after the session finishes. I ask if he remembers me.

Of course I remember you, he says. I was in your book.

I'm interested to know if you've changed your views on global warming at all, Brian. Have you been following some of the evidence coming out? Isn't it completely inescapable?

Of course I keep my views up to date, Flannery says guardedly.

How do you get your mind around all this though? I mean, aren't you even a little worried?

Aw hell, he says, grinning. He claps me on the arm and walks off.

Russia began 2006 by turning off Ukraine's gas supply in a squabble over price, hence threatening supplies to Europe. A day later, after a blizzard of international protests, they turned the pipelines back on again. Two weeks thereafter, Gazprom's deputy chairman, visiting the UK, promised there would never be a repetition of this unseemly episode. Two days after that, with an unusually deep freeze under way in Russia and a massive hike in demand for gas to keep people warm, his words proved as hollow as his pipelines.

Russia was not alone in threatening to use energy supply as an offensive weapon. In mid-January, Iran threatened to use what it called the 'oil supply weapon' if Europe and the USA continued to oppose their nuclear programme. If ever Europe and the US needed a reminder of how the world might look with oil and/or gas constrained on a permanent basis, they now had it.

January also saw a reminder that it was not just geopolitics that held the ability to constrain supply; geology could play a role too. A report appeared in *Petroleum Intelligence Week* suggesting that Kuwait had only half the reserves it had officially told the world it had. This revelation seemed instantly credible, because in December the Kuwaitis had told an OPEC summit that they would find lifting their production to four million barrels a day by 2020 impossible, unless they invited in the Western oil companies.

In February, President Bush felt the need to speak frankly of America's 'oil addiction' in his State of the Union address. He called for a programme of demand management. OPEC immediately volunteered that any reduction of demand might impact their plans to invest in new production. This was a less overt threat than Russia's and Iran's, but a threat just the same. Saudi Arabia no doubt calculated that things had moved on since the days of Jim Schlesinger.

ExxonMobil did not like this mythology about a danger from peak oil, though you would hardly call it the kind of pressure they were coming under on climate change. Having just announced the biggest profits in corporate history, courtesy of the high oil price, the world's biggest oil company now elected to spend some of its cash on an advert in the *New York Times* professing that 'Peak [oil] production is nowhere in sight . . . The theory does not match the reality.'

Based on their track record of misinformation during the climate debate, I reflected, one could surely be forgiven for concluding that their ad might be the most compelling evidence to date *for* a peak-oil problem.

Rather inconveniently for ExxonMobil, the US Army Corps of Engineers then issued a report concluding that 'world oil production is at or near its peak'. This state of affairs, the Army Corps explained, was a threat to the military, and major steps would be needed in energy efficiency and 'massive expansion' of renewables, with a move towards distributed generation, including solar PV, solar thermal, microturbines and biomass.

Such a programme of urgent clean-energy deployment, akin to the main military mobilisations of twentieth-century history, was also needed to face down the climate-change threat, in the eyes of the many concerned by it.

The catalogue of concern about climate change continued to build. February 2005 saw evidence of two of the largest glaciers in eastern Greenland doubling their speed. Melt water was evidently lubricating them. NASA climate guru Jim Hansen warned that the Greenland ice cap could collapse explosively fast at this rate, with catastrophic consequences for global sea-level rise.

Global warming now made the cover of *Time* magazine. The headline pulled no punches: 'Be worried, be very worried.'

Lower Marsh, London, 28 March 2006

David Cameron, leader of the main UK opposition party, the Conservatives, visits Solarcentury. He is on a campaign to de-toxify his party's brand, and he is draping himself and his parliamentary colleagues with as much greenery as he can find. Today's agenda is a speech about the need for a green industrial revolution. What better place to deliver it than the offices of a company trying to live that dream. As with Tony Blair's visit, the media accompanies him in some numbers.

Over the pre-speech drinks, I chat with a lady from the *Financial Times*. She wants to know which other politicians have visited Solarcentury with this kind of agenda. I explain that we now have a clean sweep of all parties. So long as they don't claim we are card-carrying members, it's perfectly fine with us. I mention that Cameron's predecessor, Iain Duncan-Smith, has also visited. He was nastily ousted to make way for Cameron, so I add that I hope my company isn't some kind of kiss-of-death for politicians.

I dare you to mention that, says the *FT* lady.

Introducing Cameron, I close by pointing to where he is standing.

Interestingly, I say, Iain Duncan-Smith once delivered a speech on the very spot that David is standing on now.

Cameron immediately takes a theatrical step to his right. The journalists, and my staff, laugh.

The man is quick on his feet, I think.

He then gives a passable speech on the green industrial revolution as a vital way to address the grave threat of climate change, and a capital way to build a modern economy.

I half listen. I have heard so very many speeches, seen so very little action.

I will get only a few distracted minutes of his time, this man who might be prime minister in a few years' time. No doubt BP and Shell could get hours with a single phone call. What do I say to him? Where do I start, among my concerns? Stray from the need to lobby for the creation of a viable solar market in the UK, and my Board will go berserk. Perhaps more importantly, I'll be letting my staff down.

It isn't easy, warning about the rumblings under the volcano. Even when you are lucky enough to get the ear of people who might be able to do something about the threat of eruptions.

Chapter 3

Doomed to failure

In April 2006, the oil price reached a new record of $72. The reason this time: the latest escalation in the war of words between the West and Iran over its nuclear ambitions. The oil industry journal *Petroleum Review* published a compilation of all oil production under way. It showed a potentially dangerous shortfall of major projects coming onstream in the years ahead. Compiling this kind of information had already, by this time, converted its editor, Chris Skrebowski, into an advocate of early peak oil.

In May, Shell announced that replacing its own reserves was no longer a forecast, but an 'aim'.

In June, Total became the first oil company to predict a year for peak-oil production. The French oil giant had long been an outlier in the oil industry, the only major to have any sympathy for the concerns of early peakers. The year it thought the peak would arrive was 2020.

Peak oil in 2020 ought to be plenty soon to cause concern, given the time it would take to mobilise renewable energy, renewable fuels and energy efficiency from the fringes of energy markets to the centre, globally. But the French oil company did not hit the front pages with its revelation.

Early peakers at the time were predicting a peak much sooner. The Association for the Study of Peak Oil's best estimate was 2010. In my 2005 *Half Gone* I elected for 2008–2012.

But all that was based on the assumption that we were through with great global recessions for a while. Few people in 2006 could see the drama that was brewing in the capital markets, and the way it would change the course of events in 2007 and 2008.

***

The Bush administration failed to torpedo the Kyoto process in Montreal. In a humiliating endgame session, in the face of a massive international outcry, it was forced to climb down. The climate negotiations were still alive.

One of the problems in the climate debate at the time was that the bad news dripping out came from one small research team or another, and as such rarely made the headline news it deserved. What was badly needed was an international pooling of scientific research and analysis such as the one that kicked the climate negotiations off in the first place, back in 1990. Then, a UN body embracing hundreds of the top climate scientists from academia and government service in dozens of countries, the Intergovernmental Panel on Climate Change, produced a consensus risk assessment. The IPCC's 1990 report, the first of three, concluded that soaring greenhouse-gas emissions would be certain to create global warming in the years ahead. It was sufficiently alarming, notwithstanding the absence of certainty about the degree of warming, and the absence of an anthropogenic footprint as of then, that it prompted governments to begin negotiating a climate treaty. This led to the signature of the Framework Convention on Climate Change at the Rio Earth Summit in 1992. That treaty had the all-important ultimate objective of 'preventing dangerous anthropogenic interference with the climate system within a time frame sufficient to allow ecosystems to adapt naturally to climate change, ensure that food production is not threatened, and enable economic development to proceed sustainably'. Developed countries agreed to adopt policies and measures that would aim to bring their emissions back to 1990 levels by the year 2000, but they agreed no targets and timetables, much less legally binding ones.

The second IPCC report, in 1995, concluded that a man-made signal could be detected in emerging patterns of global warming, and painted a sufficiently urgent picture of the implications of a world without emissions constraints that governments would feel the need to agree the Kyoto Protocol two years later.

The third IPCC report, in 2001, was more alarming still, prompting governments to keep going with the Protocol even after the US pulled out in 2001, and eventually enough governments to ratify the Kyoto treaty for it to come into force in February 2005.

A fourth IPCC report was due for international release in January 2007. There could be little doubt, for anyone following the science, that it would be even more alarming than its three predecessors. In that sense, it was a great shame that it was not ready by the time of the Montreal climate summit.

In April 2006, the Royal Society – a learned society of Britain's most eminent scientists from all disciplines – took the highly unusual step of warning that one company, ExxonMobil, would be trying hard to obscure the significance of the fourth IPCC report. Many people were surprised that such a conservative, apolitical body – usually a scrupulous avoider of any controversy – would take such a view, much less air it.

It surprised me not at all.

In May 2006, Al Gore's film *An Inconvenient Truth* made the second most money of any Memorial Day opening of any film in the history of cinema.

TV ads funded by the carbon club targeted Al Gore-type 'alarmists'. They ran an Orwellian mantra. 'Carbon dioxide: they call it pollution. We call it life.'

Lloyd's of London, the City of London based insurance market, had a different perspective. They issued a report warning that climate change could destroy the insurance industry. A world of Hurricane Katrinas, continent-wide drought-related wildfires and thousand-year floods was not one they could hope to make enough money to survive in, they said.

Vienna, 13 June 2006

A vast oil industry conference in OPEC's home city. These days, the oil and gas industry cannot gather without at least some effort to talk about climate change and the need for clean energy.

The session I am due to speak in is on the oil industry's need to get involved in renewable energy.

I go first. The usual stump speech these days: a global energy crisis is imminent; peak oil is real; your industry is in denial; the renewable energy and demand-management industries, with our 'survival technologies', will be unable to fill the gap come the crisis. This is because, I say, you have held us back throughout all the years of the great addiction. But there could be a silver lining to the cloud: maybe, just maybe, the exigencies of dealing with the global energy crisis will give the survival technologies the boost they need if we are to head off the worst excesses of climate change – as long as we are not stupid enough to resort en masse to coal and tar sands in the interim.

Talking like this to the oil industry is not the lion's-den experience it was a few years ago. They do their best, but their resistance – it increasingly strikes me – is somehow half-hearted. The first questioner, for instance, refers me to the Club of Rome, whose doom-laden predictions of resource depletion and environmental degradation so rattled the business establishment in the last century. My argument reminds him of that episode. They had been proved wrong, he says.

In this he is echoing a widely held and unfair mantra.

If that gets you to sleep at night these days, Sir, then I'm happy for you. But it doesn't work for me, or for many, many, like me. And some of us are in boardrooms.

I try not to be combative, to keep my voice calm, a small smile on my face. But with questions like this, it is hard. Don't these guys ever read about what's happening?

Next up on the platform is the man from Shell. The renewable-energy resource is vast, he says: easily enough for ten billion people even at current European per-capita energy demand. We have plenty of scope for renewable energy. We can make it economic. We have to.

The words are good, but he doesn't sound compelling.

Then comes the man from BP. He is more convincing, but he knows where the story is weak.

We will be investing $8bn in low-carbon technologies over the next ten years, he says. Jeremy would say this was a pittance, compared with our oil investments.

Too right. In the discussion, I repeat the message I have delivered to the oilmen and women so many times now, in fora like these. You haven't exactly shown the same entrepreneurial zeal on the frontiers of the solar revolution that you have for a century on the frontiers of the hydrocarbon age.

In the summer of 2006, the UK was a tinderbox of drought. So many heath and wood fires raged that public health came under threat from smoky air. Food prices jumped as a result of the miserable harvest. Europewide, waterless farmers faced crop losses measured in billions of euros. Among the casualties of the heatwave were nuclear plants on rivers, some of which had to be shut down because their cooling water was too warm.

Bad as the reality was, scientific studies pointed to a much worse future. In August 2006, the most comprehensive climate prediction to date suggested that, at 3°C of global warming, half the world's forests would be lost. In September, scientists measured five times more methane emission from Siberian permafrost than had been anticipated.

By now, those who understood the problem were fighting desperation. In the UK, climate activists tried to close down the UK's largest coal-fired power plant.

David Cameron jumped further aboard the green bandwagon by calling for the enshrining of UK carbon emission-reduction targets in law. Around this time, one could barely pick up a magazine in a bookshop without the impression that almost the whole of society was made up of environmentalists. It seems remarkable now, looking back in a world of austerity from beyond the financial crash, and amid the prolonged confusion that has reigned in the media since the incumbency's spectacular PR attack on the IPCC in the run up to the 2009 Copenhagen climate summit, to which I shall come.

Exxon came under intense pressure from its own shareholders, and a group of US senators, to change tactics on climate during 2006. In September it dropped its direct funding of the Competitive Enterprise Institute and other contrarian groups dedicated to ruining the climate negotiations.

In October, the UK government announced that it would be the first to create a law committing to carbon reductions, and the Treasury released a major report on the reasons why.

The 2006 Stern Review of climate-change risk proved to be a turning point on climate for many policymakers, because of the economic focus of the analysis.[1] Nick Stern, a former chief economist at the World Bank, led a large team who took the scientific conclusions of the 2001 IPCC report and assessed their implications for the global economy. The main finding was that the benefits of strong, early action on climate change far outweighed the costs of not acting. Without action, they calculated the overall costs of climate change to be equivalent to losing at least 5% of global gross domestic product (GDP) each year, with the clear potential for up to 20% of GDP or more. By contrast, they found that investing 1% of global GDP per annum could avoid the worst effects of climate change.

Global average temperatures rose around 0.6°C during the twentieth century, so the 2001 IPCC report concluded.[2] At around 1.5°C, the Stern Review concluded, crop yields would fall in many areas, particularly developing countries. Above 2°C water availability would decrease significantly in many areas, and a major danger would kick in: feedbacks that can lead to abrupt, irreversible changes of climate, with risk rising steadily as global average temperature increases.

Bad as this sounds, two years later Stern would be confessing that his group had underplayed the risk. They based their assessment on the third IPCC report, dating from 2001. The fourth IPCC report would not appear until four months after the Stern Review came out.

***

In August 2006, BP was forced to shut its Prudhoe Bay pipeline in Alaska after a corrosion-related leak. The leak was small, but it pushed the oil price to another record, $78.

This turned the spotlight on another element of the peak-oil risk debate: the aged state of the industry's infrastructure. Matt Simmons, the Houston oil banker who was suspicious of the state of Saudi reserves, warned that BP's pipeline corrosion could be endemic in the industry. The reason was that so much of the infrastructure – pipelines, tankers, rigs – had been procured the last time oil prices had been high, i.e. back in the early 1980s at the time of the second global oil crisis. As Simmons put it, 'the industry is kind of rusting away'. The problem could be oil's 'Pearl Harbour', he said: harbinger of $300 oil.

Worrying pointers about below-ground frailties continued to appear, for those paying attention. In July 2006, the Kuwaiti opposition to the ruling family, worried about all the rumours that there wasn't as much oil as they had thought, opposed plans to increase production. Never mind American and European SUVs, they wanted to be sure there was enough left for the use of Kuwaitis.

In Mexico, there was similar bad news. The giant Cantarell field began depleting much faster than the national oil company, Pemex, expected.

In Canada, Talisman pulled out of production in the tar sands, where costs were spiralling.

However, in September a deep-water oilfield was discovered in the Gulf of Mexico. Every time this happened the oil industry went into PR overdrive. And this was apparently a monster: as much as 15 billion barrels.

'Peak oil theory is garbage', a CERA spokesman told *Business Week* in the wake of this discovery.[3]

Peak-oil theory? Peak oil isn't theory, it's inevitability. Only the timing is in doubt. Oil is a finite resource, as every geoscientist knows. There is only so much volume of sedimentary rocks in the earth's crust capable of yielding oil.

How does one provide a counter to this? Not by writing a book and giving it to prime ministers, present and future, it seems. In September 2006, I received a letter from David Cameron saying he was looking forward to reading *Half Gone*, but his diary was pretty full, and he regretted there would be no time for a meeting to discuss it. In October I had a similar letter from Tony Blair, assuring me his government was working on the subject, and that I shouldn't be concerned.

In October, the oil price fell back to $60. 'Is the oil boom over?' *Newsweek* asked on its front cover.

Over or not, the International Energy Agency's *World Energy Outlook* for 2006, released in November, made alarming reading.[4] It issued what the *Financial Times* described as 'an apocalyptic warning'. The current energy path 'may mean skyrocketing prices or more frequent blackouts; can mean more supply disruptions, more meteorological catastrophes – or all these at the same time'. So said IEA Secretary-General Claude Mandil.

The IEA was concerned about both climate change and oil depletion, evidently. Global energy demand would surge 50% by 2030, in the case of oil requiring 116 million barrels a day, the *Outlook* concluded, with most of the increased supply having to come from Saudi Arabia, Iraq and Iran. Non-OPEC supplies were expected to peak early in the next decade, pushing the oil price as high as $130 a barrel.

Some $20 trillion of investment would be needed through to 2030 to meet this energy demand, over $4 trillion of it for oil. It was far from certain that this investment would actually occur, as things stood, Mandil warned. The apparent soaring investment by oil companies was illusory, because of inflation in drilling costs.

'This energy scenario is not only unsustainable but doomed to failure', said the Secretary-General of the IEA.[5]

Looking back in the context of the current industry excitement around the US shale gas and tight oil boom, wholly echoed by the IEA in 2012 as we shall see, these words seem remarkable. The 2006 *World Energy Outlook* does not mention shale gas or tight oil as prospects.

Cambridge University, 30 November 2006

Another debate with BP, this time in the shape of Ian Vann, Group Vice President for Oil Exploration and Production, another geologist, and a brilliant one. We face a packed lecture theatre and I am told a spillover theatre with video coverage is also full. This, then, is another surreal experience in my growing collection of such.[6]

Two weeks before, a CERA report claimed peak-oil theory was based on 'faulty analysis' and could 'distort debate'. CERA Director of Oil Industry Activity Peter Jackson wrote: 'Oil is too critical to the global economy to allow fear to replace careful analysis.' He then went on to predict an undulating plateau of production beginning beyond 2030 at well over 120 million barrels a day.

Over 120 million barrels a day.

The oil resource is actually 3.74 trillion barrels, Jackson professed, not the much lower figures peak-oil proponents claim. CERA draws on the proprietary database of its parent, IHS, for its careful analysis.

Ian Vann, taking his turn, elects to use Jackson's CERA graph showing an undulating plateau of oil supply decades into the future.

Is that what BP really believes? Even now?

But Vann has an interesting variant on the argument, and one I haven't heard before. He seems to be agreeing with both the CERA and IEA assessments.

The peak-oil debate is not about geology, he says. The total oil endowment is probably more than five trillion barrels, if we include the tar sands and all the other unconventional oil resources.

The CERA view, on steroids. He is including the Green River Formation in Colorado, Utah and Wyoming in this – an oil shale deposit which contains two trillion barrels or so of a deposit wherein the hydrocarbon takes the form of kerogen, and from which there is no known process to extract oil at usable scale.[7]

So what is it about then?

It's about economic irrationality, Vann tells the assembled professors and students of Cambridge University. If decisions continue to be played out as they are today, in circumstances of economic irrationality, for the protection of the state and in many instances of national companies that are not open – and in many cases are not even competent to do the job – then oil supply will become very difficult.

The possible consequences, he adds, are horrifying.

As he speaks, I watch the audience, imagining them as an oversize jury charged with discerning truth in the post-mortem beyond peak panic about collapsing global oil supply some years from now. BP's case, it seems, will be that they were right about peak oil all along, and that it was geopolitics that prevented them from getting all the oil out.

There is 40 years of supply in those reserves, they'll say. And look at all those resources. We just couldn't get at enough of the reserves, and we weren't allowed to turn enough of the resources into reserves.

And we early peakists will say peak oil is all about flow rates, it's about both below-ground factors and above-ground factors; BP and the rest knew this all along, and are obfuscating.

But few will care about bickering forensics. Because the crash will be upon us.

Chapter 4

We are not responsible

The year 2006 had begun with Moscow turning the taps off on a neighbour in order to encourage higher payments for gas. So too did 2007. This time the target was Belarus. Belarus immediately slapped a transit tax on Russian oil exports through the pipeline crossing their country from Russia to the West. About 40% of Russia's exports then crossed Belarus in the inappropriately named Friendship Pipeline, carrying 1.2 million barrels of oil a day into Europe, 12% of European imports. The pipeline provided 20% of Germany's needs and 96% of Poland's. The Kremlin duly turned the taps off. Belarus blinked first in the stand-off and cancelled the transit tax on Russian oil.

The EU then roundly threatened Russia with the loss of energy contracts if they turned off oil and gas pipelines ever again. The EU at the time imported a quarter of its oil and 42% of its gas from Russia. The Energy Commissioner, Adris Piebalgs, called for accelerated EU investment in renewables. It would have been clear to him and many others by now that they could pose all they liked, but the Putin regime would not hesitate to use oil and gas supply as a weapon exactly when and where it saw fit.

Meanwhile, new fears emerged about supplies of both oil and gas, this time from Iran. Iran supposedly had the third largest oil reserves in the world at the time, after Saudi Arabia and Russia, and the second largest gas reserves, after Qatar. But chronic underinvestment and soaring domestic demand for oil, driven by the lowest prices paid at the pump anywhere, looked as though they were squeezing the proportion of oil available for export. One analyst forecasted zero oil exports as soon as 2015.[1] The Economist Intelligence Unit reported that the Iranians were using so much gas to boost oil production that gas exports were falling under threat too.

At the annual Detroit motor show in January, it became clear that the persistently high oil price had plunged US automakers into turmoil. The cost of gasoline and steel had also gone through the roof. Their customers were switching to smaller, greener vehicles. Hybrid sales had jumped sixfold over four years. The Hummer stand at the show was thinly populated.

GM unveiled another electric car at the show. Their first, many environmentalists suggest, had been deliberately killed to hold the technology back. Now, in a seismic reverse, GM said it hoped the new electric car would spark the giant corporation's revival. The Volt, a concept car that could do 40 miles per charge with a plug-in battery plus onboard recharger, could run off either gasoline or a fuel cell.

***

The 11 warmest years, in terms of global average temperature, had all been in the last 12, and the broken records continued in 2007. January was the hottest since records began in 1850. New Yorkers basked in 22°C heat. In Britain, after the hottest year ever in 2006, the daffodils were out. In Moscow, the zoo's bears couldn't hibernate because the weather was too warm. The thermometer read 5.3°C when it should have recorded minus 18°C.

The European Commission called in January for a 20% collective cut in greenhouse gases and 30% globally by 2020, saying that unmitgated climate change would devastate the Union. Unveiling its energy strategy, the Commission said this combination of targets should get the world on track to keep below 2°C of global warming, if others joined in.

Corporations were changing their minds on climate change in growing numbers. Ten CEOs of US companies called on President Bush to support carbon dioxide cuts of 60% by 2050. Forming the US Climate Action Partnership (USCAP), Alcoa, BP America, Duke Energy, DuPont, Caterpillar, General Electric, Lehman Brothers, FPL Group and PG&E reversed an opposition to carbon regulation that had once seemed monolithic in American industry.

Radical corporate action was not limited to calls for emission controls. The giant Allstate Insurance company pulled out of Delaware, predicting ruinous weather events as a result of global warming.

Corporate America was finding its normal programmes of greenhouse denial increasingly difficult to run. The Union of Concerned Scientists published a report alleging that Exxon paid $61 million to 43 climate-change denier organisations between 1998 and 2005. The report concluded that ExxonMobil had been the worst offender in this kind of subterfuge. Another report exposed systematic attempts by the Bush administration to doctor scientific evidence on climate. This survey of 1,600 government scientists showed that fully 46% of them had been warned against using terms like global warming in their reports. Phil Cooney, a big player in this campaign in the White House, now worked for ExxonMobil.

Under pressure, Exxon cut its ties to greenhouse sceptics, including the Competitive Enterprise Institute, and agreed to meet with non-government organisations to discuss global warming.

George Bush hadn't mentioned climate change in his last five State of the Union addresses. But in January 2007 he was forced finally to address the subject. The best he could do was propose a 20% cut in gasoline use within a decade.

The concern about global warming that had been evident for over a year reached a peak on 8 February with the release of the Fourth Intergovernmental Panel on Climate Change Scientific Assessment Report. The worldwide headlines the next day had a clear theme: worse than we thought.

Three hundred delegates representing 600 scientists from 113 countries attended the final IPCC drafting meeting. Some 2,500 scientists in all collaborated on the report. They concluded that global average temperature rise will 'most likely' reach 4°C by 2100, and possibly up to 6.4°C, in the absence of cuts in emissions. They had concluded 5.8°C at the most in their 2001 report. In 2001, the conclusion that warming was man-made had been deemed by consensus 'likely'. Now it was deemed 'very likely', and that would have been 'virtually certain' but for the interventions of China and a few others in the final drafting session.[2]

In 1990, working as a scientist for Greenpeace, I had listened to the experts who had completed the IPCC's first assessment in a Berkshire hotel. At a press conference, Margaret Thatcher, not otherwise known for eco-doom-mongering, warned the report would 'change our way of life', and that we would cry out in the future not for oil, but water. The world seemed to be listening. The UN called for multilateral negotiations and most governments signed up.

But these had run now for nearly a quarter of a century, and done little to stem greenhouse-gas emissions.

In 1990, I watched dozens of lobbyists from Exxon, OPEC and the world's coal groups try to persuade the IPCC's scientists to soften their language. They failed. Over the years since, the warnings have only strengthened, despite the incumbency's best efforts.

In February 2007, the Intergovernmental Panel's fourth warning was all across the front pages and TV news the morning after its publication. Even the US networks turned up in force to the scientists' news conference in Paris. Between the committee-written words, you could smell the panic that existed behind the scenes. For the *Guardian*'s man on the spot, the graphs said it all. 'The words "hell" and "handcart" came to mind', he wrote.

In early February, an Exxon-funded lobby group, the American Enterprise Institute, offered $10,000 cash to scientists who would critique the report. It seemed to many observers, early in 2007, that the last bastions of neoconservative climate denial were becoming desperate.

Now, looking back, it seems increasingly likely that no amount of evidence will ever be persuasive for hard-core contrarians. In due course they were to find a way of turning the tide. And events in the capital markets were to help their cause in a very big way.

Hong Kong, 2 March 2007

On the top floor of a skyscraper I sit for more than an hour talking about climate change with Shell's CEO Jeroen van der Veer and two other chief executives generally favourable to his world view. The debate will be screened later on BBC World, but is to be filmed as though it is live.

My role is to be the attack dog. Be as pushy as you like, the presenter has told me. But you are forbidden from mentioning peak oil. This is about climate change.

I open with the obvious point. The most recent IPCC scientific assessment points essentially to the imperative of a managed withdrawal from fossil fuels over time. The Greenland ice cap makes the case. If we burn enough fossil fuel to go above 450 parts per million of CO_2 in the atmosphere, we now know we face the grave risk of locking the Greenland ice sheet into a total meltdown. If we did that, global sea levels would go up by as much as seven metres. Much of the world economy is based on coastal plains.

Just look around.

I wave my arm across the smoggy Hong Kong skyline, visible through the windows.

We would be on course essentially to lose the world economy. Are we really on the right track to stop the melting of the Greenland ice sheet, do you think?

If you say withdraw from fossil fuels, I disagree, says van der Veer firmly. There are ways to deal with carbon emissions. Carbon sequestration, for example. The art is to make the fossil fuels green: to make the CO_2 go away. You can do useful things while making this happen, such as using the gas pumped underground to enhance oil recovery.

I'm not saying we shouldn't use sequestration, I reply. It's a tool we'll have to use, among many. But the point is that Shell is planning huge new emissions: new oil production in the Arctic, the squeezing of oil from the Canadian tar sands.

Beyond all this, Shell is working with the Chinese on ways to distil oil from coal. We've only got ten years to turn this problem around and get to deep cuts. How are we going to do that given Shell's plans for so many new emissions? As for sequestration, the UK is planning a programme at present that involves only trials, and would not allow any mass production of sequestration technology before 2012. Too late.

In mid-February, van der Veer had given a speech during International Petroleum Week. The IPCC's final warning had come just two weeks earlier. He had offered a vision of growing oil use and a coal comeback. He had said the most important fact was that energy demand was rising and would continue to rise. Another fact was that fossil fuels were and would remain the dominant source of energy for decades to come.

Maybe, he had argued, we could cut fossil fuels in the energy mix to 77% by 2030 if we tried hard. Renewables perhaps could be 25% of the energy mix by 2050.

I sit trying not to stew, endeavouring to think fast through a treacle of jetlag and recent sleep deprivation, trying to get the balance right between being thought-provoking and combatively rhetorical. With the energy system van der Veer envisages for 2050, we wouldn't have a chance. I want to bring this point out badly.

Van der Veer is picking his words carefully, almost hesitatingly.

To a certain extent I agree with you, he says. But if you really would like to make fast reductions, you need co-ordination of standards.

I am not clear what he means.

Governments are not going to agree standards, I reply. US administrations had done a good job of messing that possibility up. The ball has been thrown to business. It's up to us to recognise that if we do chase all these new and unconventional oil and coal projects around the world, we can't hope to stay below the danger threshold of 450 parts per million of CO_2 in the atmosphere.

Jeroen van der Veer is now clearer in his response.

We need energy. If a country decides that we need to develop our oil sands, like Canada, why should Shell say we don't go there?

Because to do so would involve you cooking the planet and killing your customers, I think of saying. But I don't. Shell has recently hired three Bush administration officials to help open up oil shales and Arctic conventional oil. Should I try and score a cheap point on that one?

Because BP has said that it won't exploit the tar sands, and for climate reasons, I choose to say. BP under your equivalent, John Browne, takes the view that you can't go for the tar sands and be serious about climate change. Do you disagree?

The Shell CEO sidesteps.

But we started investing in solar and wind eight years ago, he says.

I choose to ignore this red herring about Shell's half-hearted renewables effort.

I think we are talking about the dysfunctional, nay ... suicidal, heart of the machine here, I say. I bet if you, Jeroen, woke up and decided that in fact we would go over 450 parts per million if Shell went to the Arctic and the tar sands and all the rest, and you came out and said it, you'd have a terrible time with your investors.

Now van der Veer comes to his bottom-line argument.

The way I view it, I'm an enterprise, I'm not a government. We develop products which are best of their class. Shell is not ultimately responsible for the energy used in the world.[3]

I shut up, hoping at this point that his words speak for themselves.

There you have it, I think. Shell is not responsible for the energy used in the world.

The drug pusher's argument.

Looking back, it becomes clear how much the oil industry has dug in since then. Shell never did any meaningful carbon capture and storage. Van der Veer's art of turning fossil fuels green turned out to be just artifice. They pulled out of their UK effort, citing excessive costs. As for renewables, Shell pulled out of solar completely.

And I was too quick to hold BP up as a paragon of virtue in the tar sands. CEO Lord Browne was forced to step down in May 2007, having presided over one exploding oil refinery and pipeline leak too many as a result of his cost cutting. Tony Hayward replaced him. BP abandoned its pledge not to exploit the tar sands before the year was out. They pulled out of solar completely in December 2012, just as the technology began to approach price parity with conventional electricity in multiple countries.

On the banned subject of peak oil, within a few weeks of the BBC World debate the US Government Accountability Office announced that America needed a plan. This investigative arm of Congress noted that peak-oil forecasts ranged between tomorrow and 2040. The consequences of a peak and permanent decline in oil production could be even more prolonged and severe than those of the past global oil supply shocks, in the late 1970s and early 1980s, they concluded. There was no formal strategy for co-ordinating and prioritising federal efforts to deal with peak oil, and there should be.

Peak oil and climate change were now coming to be seen as threats to national security by an expanding constituency, on both sides of the Atlantic. In April 2007, 11 former US Army generals announced that they had come to the conclusion that climate change was a significant threat to US national security.

And almost unnoticed at the time, a sub-prime mortgage lender called New Century Finance filed for Chapter 11 bankruptcy protection.

Chapter 5

The risk of contingency

Brixton, London, 4 June 2007

A town hall in a deprived borough. An evening talk to a chapter of a new social movement. So-called Transition Town groups are springing up around Britain. A new umbrella organisation, the Transition Network, co-ordinates them. This is my first invitation to speak to a Transition group. I am intrigued.

In the hall sit around a hundred citizens: a completely random mix of any hundred you might see on a London street. Quite a few are in suits, obviously having come straight from work.

What makes them give up an evening? Concern about peak oil. They have gravitated to an emerging philosophy that believes local efforts to live less oil-dependent lifestyles will be essential both to help governments tackle the peak-oil threat and to prepare for the aftermath of the politicians' probable failure to head off an oil crash. Many of them are also worried about climate change, and know that the right responses to the peak-oil threat can also buy society the emissions reductions that help reduce global warming.

For these reasons, they are folk interested in resilience: the many ways that people and communities can help themselves survive and eventually prosper in an oil-constrained world. They organise their group, like all the other Transition Network groups, around projects such as allotment-based vegetable cultivation, community renewable energy, and the creation of a local barter-based currency.

I give my talk, an update on the issues that concern me and them both. The questions afterwards are perceptive and deep. They come thick and fast.

I see that the group is apolitical. When local politics comes up, it is clear there are voters of all persuasions attending. What strikes me most is the spread of vocations: teachers, architects, shop owners, local government workers, and so on. Over tea and biscuits afterwards, one tells me that he is a civil servant. Another that he is a policeman.

A broad spectrum of the citizenry not only disbelieves the incumbency's narrative on oil, it seems, but is also scared enough by it to act unilaterally.

St James's Palace, London, 21 June 2007

The Labour government holds a forum on climate change. Al Gore is in town for it. A hundred-plus business leaders sit in tables of eight, listening to him and Chancellor Gordon Brown, who takes over from Tony Blair as prime minister tomorrow.

Between speeches, the tables become discussion groups, each chaired by a senior civil servant. Brown and his Foreign Secretary, David Miliband, tour the tables, listening in.

One discussion is on the theme of mobilising society to deal with climate change. I just begin talking to my group about the emerging phenomenon of the Transition Network when Miliband sits down at the table. He and I are old sparring partners over the Labour government's performance on climate policy. Earlier, walking in, he had stopped to say to me, with a grin: Good to see they've got you at the back, Jeremy.

He listens for a while and moves on.

Later he returns with Gordon Brown.

Jeremy, he says, Gordon wants to know what a Transition Town is.

I tell him, trying to ignore the frustration on the faces of the business people round the table. It is a wonderful pure-chance opportunity to deliver the peak-oil message to a head of government, sideways as it were. I tell

Brown I will write him a letter explaining more. It's not just the Transition movement that is worried about this threat, I make clear.

Al Gore gives the closing speech. Governments mustn't wait until 2012 to make the next big commitments on climate, he says. We must negotiate a new treaty by 2010. The Montreal Protocol on ozone depletion serves as a precedent. It was negotiated in 1989 in Montreal. It was strengthened in 1990 in London. Business came on side to make that happen. We need the voice of progressive business, but we also need political leaders who follow the shining example set by Churchill. His use of words, his passion and wisdom, caught the imagination of the people and motivated them to do extraordinary things.

Sad amateur occasional historian that I am, I own a full set of Churchill's wartime speeches.

We shall fight on the beaches, we shall fight on the landing grounds, we shall fight in the fields and in the streets, we shall fight in the hills; we shall never surrender.

What would the climate equivalent have been, had the old boy been prime minister today? The enemy was so clear then.

I sit watching the two leaders go about the theatre required of them in their jobs: the man who was nearly president of the United States, the man who will be prime minister of the United Kingdom tomorrow, both of them fighting a certain intrinsic awkwardness in their personalities. They both care deeply about the threat of climate change. That much I know. Both understand how gravely it threatens civilisation. Both have called the battle to abate it a moral imperative.

I look around the room. The usual suspects are all here, or their adjutants are: the corporate chiefs who will turn up to an event like this and nod at the right points, make the right noises, then go home and do little that deviates from the status quo, meanwhile letting their heads of lobbying protect the corporate interest however they think best, without having to answer too many questions of detail.

And all the while they will not feel too bad about themselves because, when all is said and done, they are not ultimately responsible for the energy used in the world.

In June and July 2007, the International Energy Agency blew another very loud whistle on oil depletion. 'World will face oil supply crunch within five years' read the front-page headline in the *Financial Times*. The IEA's Mid-term Market Report now foresaw very tight global supply in 2012. Much would hinge on whether production could be lifted in Iraq, Chief Economist Fatih Birol said.[1] The IEA was also predicting major problems with gas supply before the end of the decade.

Let me emphasise this. At this point in the history we have reached 2007. By 2013, the IEA would be back on the bandstand with Big Oil, prophesying oil and gas aplenty, and even the prospect of an energy-independent America. I will return to this remarkable turnaround when we consider the shale gas boom from the vantage point of 2013. At that point I shall ask the obvious question. Has the risk really receded that much?

In July 2007, the US National Petroleum Council (NPC) conceded that there were indeed 'accumulating risks' to oil and gas supply. Surprising as the IEA change of heart since 2005 had been to early peakists, the NPC statement was astonishing, since this was a body chaired by ExxonMobil CEO Lee Raymond.

'The world is not running out of energy resources, but there are accumulating risks to continuing expansion of oil and natural gas production from the conventional sources relied upon historically', the NPC said. These risks 'create significant challenges to meeting projected energy demand'. Especially when a big oil project can take 15–20 years between exploration success and first production.[2]

In August 2007, Conoco's CEO Jim Mulva warned of 'serious future gas shortages', saying 'the world has a natural gas problem'.

In July 2007, the oil price soared to a giddy $79, and nations jostled for position in the race for the diminishing supplies. Canada warned other nations to keep their hands off its Arctic oil. Russia planted a flag on the sea floor four kilometres below the ice at the North Pole. (The Kremlin's underwater footage, released to national TV networks and duly broadcast to the world, was later observed by a teenager to have been borrowed from the film *Titanic*.)

The White House urged OPEC to lift its production. As a wag observed at the time, 'How dare they be so slow in producing our oil.'

The Saudis began recruiting a 35,000 strong oilfield protection force. Perhaps they feared a recurrence of Schlesinger-type thinking in this kind of climate.

The oil price crossed $80 for the first time in September 2007, and $90 in October.

The petrodollars flowing into the OPEC states were now bankrolling massive expansion of infrastructure. A $600 billion build programme was under way in Saudi Arabia. This, of course, would only serve to worsen the oil supply problem down the track, forcing domestic oil consumption ever higher as it did.

Releasing the IEA's *World Energy Outlook* in November 2007, Fatih Birol warned that the world had only ten years to turn round its energy policy. The IEA's message to the industrialised world was stark. 'There is a need for an electroshock. We have to act immediately and boldly', Birol said. In the next year, China on current trends would install 800 gigawatts of power-generating capacity, around as much as Europe now has. The implications for coal burning and greenhouse-gas emissions did not bear thinking about. And to meet projected oil demand by 2030, OPEC would have to double supply.

Sadad al-Husseini, the man who knew the old Saudi oilfields better than anyone, scoffed at that prospect. In October 2007, he professed that global peak oil had already arrived, in terms of crude oil. 'We are already three years into level production', he said.

At this point let me introduce a little detail about different types of oil, for those unfamiliar with oil markets. Most of what we use today is crude oil – the type of oil that we see on beaches when tankers hit the rocks. This is the easy stuff to extract. To this we have to add small amounts of condensate and other natural gas liquids, components of natural gas that are liquid when the gas is produced. At the time of writing in April 2013, global production of crude has been on a plateau of around 74 million barrels a day since 2005. Crude-plus-condensate-plus natural gas liquids have been on a plateau of 82 million barrels.[3] This is what al-Husseini was referring to in October 2007. That plateau has not changed since. What has changed is that total production in 2006 averaged 85.2 million barrels a day and in 2011 averaged 87.4 million barrels a day. The rest of what is referred to as 'All Liquids' production is made up of refinery gains, unconventional oil – from tar sands in Canada and latterly from tight oil – biofuels and oil released from stockpiles. I will return to these plateaus in crude and crude-plus-condensate-plus natural gas liquids and their implications, but for the moment let me simply flag the numbers involved here. It surprises many people, in the face of all the hype, to find that crude production has been on a plateau since 2005.

The oil price stood at $98 in November 2007. $100 oil was now just a matter of time. Interestingly, given the whiffs of economic panic that had accompanied mere $40 oil, the $80 and $90 landmarks during 2007 had not merited front page news.

The reason, perhaps, was that by this time the relevant people had other, more immediate, things on their minds.

***

By June 2007, a sense of grave concern was building in the markets over the mountain of sub-prime mortgage debt in America. In July, a Bear Stearns hedge fund investing in sub-prime mortgages collapsed. The chairman of the US Federal Reserve warned that sub-prime defaults could top $100 billion.

The key figures supposed to regulate this kind of risk had seen so little of it coming, and even now were woefully short of appreciating what was really under way.

On 9 August 2007, the credit markets froze. Fearful of the losses that their peers might have accumulated, banks lost all confidence and stopped lending to each other. Central banks poured $323 billion into the world economy to try and restore confidence. This sum, equivalent to a quarter of the UK's entire annual output, released over just 48 hours, did not shift the problem.

As nerves failed in ever greater numbers, the credit crunch in turn triggered a spending squeeze. On 13 September, the UK suffered the first run on a bank for decades. Northern Rock, the UK's fifth biggest mortgage lender, had to be bailed out by the Bank of England.

The inquests began on multiple fronts. It would only be a matter of time before the heads began to roll. Merrill Lynch's CEO was the first to go, followed soon after by Citigroup's. As recently as July, this man, Chuck Prince, had justified his bank's involvement in the sub-prime debacle as follows: 'When the music stops, in terms of liquidity, things will be complicated. But as long as the music is playing, you've got to get up and dance. We're still dancing.'[4]

Within weeks, the dancers were fleeing the dance floor in droves, leaving carnage behind them.

Whitehall, London, 29 November 2007

A room in the Department of Business Enterprise and Regulatory Reform, the ministry responsible for energy. The department's Chief Economist has agreed to discuss peak oil with me.

My book on peak oil, *Half Gone,* has come and gone, in the manner of many such books. Two years after its publication, kind souls have been suggesting to me that the peak-oil debate has moved on so much – in the general favour of the arguments in my book – that a follow up is required. But I think I have a better idea. The world is full of middle-aged men with a bee in their bonnet and a book about it. If this idea is to be taken seriously, it will be far better made by a taskforce of companies, using business risk arguments. I have resolved to try to put such a group together.

Virgin is up for it. SSE one of the Big Six UK energy companies and an investor in my company Solarcentury, is also on board. Others are considering. We are taking the obvious view that it would make much more sense if the taskforce were a joint industry and government body. After all, if there are indeed reasons to fear early peak oil, then national security enters the frame. I am charged with putting the case to government. This is the meeting where I will do so.

It does not start well. The Chief Economist tells me that the oil price will continue to go up and that will furnish the oil companies with more cash for exploration. So they will explore for oil more, and find more oil.

This is the classic economist's argument, when it comes to peak oil.

I ask her where she has seen this happen in recent decades.

I have seen a chart in a report, she says. I'll look it up for you.

I swallow, change tack.

Look, I say, even the IEA is now saying that we can forget non-OPEC oil production rising. Are you happy to put all our eggs in the OPEC basket?

Even if the IEA is right, she says, we have every reason to believe that OPEC has enough oil to cover the gap.

She delivers her answers, it strikes me, as though this is all a role play exercise with which she can't quite engage seriously.

I run through the reasons why not, country by country, for Saudi Arabia, Iraq, Iran and Kuwait. Underinvestment is a big theme.

They are not investing in new production because they are happy with near-$100 oil, she says.

Now I am confused. She just told me that the oil industry *can* cover the depletion gap if the oil price is high.

She has two other junior officials with her, a man and a woman. Junior man chips in at this point.

Even if there is a problem with crude oil, he says, the tar sands can come to our rescue. There are more than a trillion barrels of oil resource in the tar sands.

Yes, I say. In a solid state. Requiring the burning of a vast quantity of gas to heat the water to melt just a little of the tar. Ignoring the climate-change implications of that ghastly process, what is the government's assessment of the flow rates possible from tar sands by 2015?

I direct my words to the Chief Economist.

She shrugs, and looks at her junior officials.

It strikes me that it is as though she is saying 'I am the Chief Economist. I don't do detail.' And she seems perfectly comfortable with that. Indeed, she has an almost couldn't-care-less air.

Junior man and junior woman look at each other.

That is the point I realise that these people, who are responsible on the government side for a large slice of the nation's energy policy, probably have no idea to within an order of magnitude what the flow rates from the tar sands will be. They just 'know' that there are a trillion barrels of oil in the tar sands.

Three million barrels a day, I say. In a world where even ExxonMobil says the old crude fields are depleting at four to five million barrels a day.

Oh Dr Leggett, says junior man, that's your figure.

He smirks, clearly thinking he has scored a point in the eyes of his boss.

No, I say. That's the Canadian oil industry's figure.

I manage to say it calmly. But inside I am boiling like a tar-sand melting plant.

A pause. And so to the bottom line.

Even if you don't agree about a premature peak, surely you must accept that there is a *risk* issue, I ask, trying to avoid an imploring tone. In recent

weeks people like Saudi Aramco's former head of exploration and production and Total's current CEO have spoken out about the danger of an
early production peak. Why not at least join with the UK companies studying
peak-oil risk and draw up a contingency plan? Why is that such a big thing
to ask?

No, says the Chief Economist, confidently and without hesitation. A
contingency study would be too risky. The existence of a government study,
she explains, might be leaked, frighten the market and create needless panic.

Chapter 6

The small print

Two days into 2008 oil hit treble figures for the first time. 'Oil at $100 threatens to choke the economy', shouted the front-page headline in *The Times*.

But who to believe on peak oil, and the risk that choking would turn to something even worse? In January, Total warned again that in their view peak production was near. CERA put out another study saying it was far off.

A James Baker Institute analysis sat uncomfortably with CERA's bullish position. It urged oil chiefs to address falling investment in exploration. The big five international oil companies had cut exploration spending in real terms between 1998 and 2006, in spite of the rise in oil prices. ExxonMobil, BP, Chevron, Royal Dutch Shell and ConocoPhillips used more than half (56%) of their increased operating cash flow not on exploration but on share buybacks and dividends.

Other oil bosses joined Total in the pessimistic camp. Admitting the industry was having trouble finding big new oilfields, Chevron CEO David O'Reilly talked of a future for his company in energy services. 'We're a pretty resilient bunch', he said in an interview with *The Chronicle*. 'We'll be around. We'll be selling energy. We'll be providing energy services. But I'm confident it will be quite different than it is today.'[1]

In February 2008, John Hess, chairman and chief executive of Hess Corporation, told the CERA annual meeting in Houston that the industry was heading for an oil crisis within the next ten years. Some 60% of the world's oil production was from countries that had already peaked, he said. Tar sands needed to be encouraged, but their contributions to supply would not be material enough to bridge the gap in oil requirements over the next ten years.[2]

Production had fallen at ExxonMobil, Shell and BP in 2007 despite $60 billion in capital expenditure. Shell's tar sands production was falling. Foreshadowing what was to come in production from shale, the company pleaded with the SEC to be allowed to book unconventional oil and gas as reserves.

History shows us, then, that the industry has shifted from special pleading about reserves accounting to a message that they have an excess of supply in five years at most. At the same time as this special pleading was going on, some UK Industry Taskforce members were expressing surprise at discovering what early peakists had long known: that the average time between the discovery of a conventional oilfield and first production was six to seven years.

Demand projections, meanwhile, looked increasingly problematic. In India, a car costing a mere £1,290 went on sale. In China, jet fuel demand seemed set to grow by 11–13% each year until 2020.

In March, the IEA decided to up the alarm. 'Oil production by public companies is reaching its peak', Chief Economist Fatih Birol announced. 'They will have to find new ways to conduct business. Increasingly, output levels will be set by a very few countries in the Middle East. Tar sands are attractive, but, like biofuels, they will never replace Middle East oil. In the long term, we must come up with an alternative form of transport, possibly electric cars, with the electricity being provided by nuclear power stations. The really important thing is that even though we are not yet running out of oil, we are running out of time.'[3]

No mention of shale gas. No mention of shale oil. Not a hint, indeed, of the message of plenty to come just a few years later.

The next day, with fear in the air that Saudi Arabia would not lift production, oil broke the all-time inflation-adjusted price record of nearly $104.

Russia, of which so much had to be expected in terms of new production, was looking increasingly unreliable. News from Moscow was dominated by Kremlin pressure on BP, whose hopes of stability and new reserves via their joint venture TNK-BP were evaporating. In February, a Russian senator voiced doubts that Russia could meet its energy commitments to the West. By April, oil companies brave enough to run the gauntlet in Russia were encountering a new problem: unavailability of debt for development funding as the credit crunch bit ever deeper into bank lending.

In May, George Bush visited Saudi Arabia and appealed directly to King Abdullah to lift production. The King agreed. But less than a week after this, the IEA issued yet another warning about an early oil crunch. Its new study of supply, to be published towards the end of the year, was the first to look field-by-field around the producing countries, the agency said. It was not finding good news.

Those following the peak-oil drama lost no time in pointing out the obvious. Why on earth had it taken until now for the IEA to do a global analysis field-by-field?

Oil duly passed $135, breaking records for three days running, up an unprecedented $10 in a single week.

In June it went up $10 in a single day.

With the oil price near $140, the IEA announced on 10 June 2008 that the world officially faced an oil crisis.

'We can call it an "oil crisis" given the current price, and that it continues to climb even after global efforts to cut consumption', IEA Secretary-General Nobuo Tanaka said. 'We see a critical, structural issue in the global oil market, where supply growth isn't catching up with demand.'[4]

BP headquarters, St James's Square, London, 11 June 2008

The publication of BP's annual Statistical Review of World Energy is particularly timely this year. The energy commentariat are assembled to hear what this digest of proved reserves, production and consumption has to say.

Tony Hayward has written an op-ed in the *Financial Times* calling the Review one of the most reliable sources for energy data worldwide. It is certainly one of the most quoted. Everyone interested in energy uses it: politicians, policy wonks, students, journalists.

But how reliable is it really?

Tiny printing on the inside cover of the document reveals a catch-all caveat. The information presented comes not from primary BP research, it reads, but from 'official sources and third-party data' and 'does not necessarily represent BP's view of proved reserves by country'.

This astonishing get-out clause has been inserted in the BP review every year since Shell was caught lying about its reserves in 2004.

In the Q&A session, I stick my hand up. I see small grimaces appear on faces as I speak. I am not among friends. This is an audience that laughs politely every time BP speakers tell what I have come to think of as the annual peak-oil joke.

As BP Chief Economist Christoph Ruhl likes to say, we at BP don't believe in peak oil.

Not 'we don't believe in early peak oil'. We don't believe in peak oil.

I ask BP's lead author whether he can put an uncertainty range on the data in the report, since it comes with a health warning in the small print.

No, he says, I am not able to do that. Certainly, he concedes, there are some good apples and some bad apples in among the data from the official sources and the third parties.

Those of us who worry about peak oil know about a few of these bad apples. That is one of the many reasons we are worried.

OPEC decided to set its annual production quotas according to size of national reserves in 1983. A few years later Gulf countries started finding that they had underestimated their reserves. They added more than 300 billion barrels to their collective tally: fully a quarter of current supposedly proven global reserves of 1,200 billion barrels. From then on – year after year, country by country – they have tended to report the same figure for proven reserves as they did the year before. They ask us to believe by strange coincidence that they find exactly the same amount of oil each year that they sell to the world market.

And BP relays it all in its annual review, claiming the document is one of the most reliable sources for energy data worldwide.

Except in the small print.[5]

During June 2008, the inevitable protests began against high fuel prices. In Spain and Portugal, supermarket shelves stood empty as blockades cut deliveries. In France, haulage unions planned motorway gridlock. In the UK, a fuel strike was in the offing. It threatened only one in ten petrol stations, but panic buying started immediately. In rural Britain, thieves took to draining farmers' diesel tanks.

It is in small episodes such as this, here one day and gone a few days later, that we are given chances to appreciate the extent of the challenges society will face, should a sudden drop or even collapse in national oil supplies descend on us.

Renewed tension between Iran and Israel drove the price above $144 at the end of June. On Nymex, investors bet for the first time on $300 oil before year end.

***

The credit crunch was meanwhile unfolding in a steady drumbeat of economic pain, dampening the urgent IEA messaging around the oil crisis. In February 2008, G7 finance ministers warned that sub-prime losses would exceed $400 billion. The UK government felt compelled to nationalise Northern Rock. In March, the crisis claimed its first American bank. Bear Stearns, the fifth biggest investment bank, was sold for just $230 million. By May, fears of stagflation were growing: 'An economic serial killer is on the loose', as the *Economist* put it.

The hardship was tough to watch. In Spain, the construction industry all but collapsed. Growing numbers of Americans were being forced to live in their cars. Less than a year after the crisis broke, structural change was already evident in the US. Language was even changing with it.

'Reurbanisers' had begun leaving the 'ghostburbs' of America.

***

In my account of energy-risk history thus far, I have made little reference to clean energy. But ever since the first upward jolt of the oil price in 2004, a partial success story had been building quietly. Governments concerned about both climate change and energy security had been introducing market stimulation measures of various kinds, and by the time of the credit crunch this had driven impressive growth in renewable energy and renewable fuel industries. By 2007, global investment in renewables totalled nearly $150 billion, up from $70 billion in 2006. The number and value of renewables companies were soaring, and big companies were showing signs of smelling the green coffee. GE, for example, announced in January 2008 that a quarter of all its energy investments would be in renewable energy by 2010. More than a thousand private equity funds were active in clean energy.

In Silicon Valley, a 'cleantech' revolution was being widely pushed as the natural successor to the internet revolution. The world's biggest venture capital investor in cleantech, Vantage Point, invested in my company Solarcentury in 2006, so I was visiting the Valley regularly and seeing this phenomenon close to.

In April 2007, Guinness Asset Management went so far as to run news-paper advertisements both promoting its alternative energy fund and writing an obituary for fossil fuels.

Two things were inevitable, though, with the benefit of hindsight. First, the credit crunch would harm this process. Fast-growing renewables companies would be needing credit for working capital, and many banks had essentially stopped lending after August 2007. Second, elements of Big Energy would not be culturally disposed to assisting in any talk of revolution. BP and Shell were by this time overtly recarbonising. By May 2008, when Shell ditched its share of the UK's largest wind farm, talk had begun of a global retreat from renewables.

It was not going to be a retreat, but renewable energy was going to be a very tough space to operate in from here on. The incumbency was in the process of turning the energy markets into a civil war zone.

St James's Palace, 16 July 2008

The Prince of Wales is not among those who seem to think a credit crunch is reason for forgetting the slower-moving crisis in the climate system. His Cambridge Programme for Sustainability Leadership has gathered together a hundred-plus business executives to be updated on the very latest science, economics and politics of climate change. The aim is for our companies to sign a Business Declaration on Climate Change urging action at this year's climate summit, to be held in December, in Poland. Just as we did for the 2007 summit in Bali.

Professor Hans Joachim Schellnhuber, head of the Potsdam Institute and a leading light in the Intergovernmental Panel on Climate Change, summarises the latest science. He now talks of the need to return atmospheric con-centrations of carbon dioxide to 280 parts per million. A common assump-tion up to now has been that 450 ppm might give society a good chance of staying below 2°C of global warming since pre-industrial times. The reason for the seemingly extreme new target, Schellnhuber explains, is the long-run sea-level risk.

Think of a 1°C global temperature increase as locking in 20 metres of long-term sea-level rise, he says. That would be apocalyptic enough, but of course the higher greenhouse-gas concentrations go, the more we risk awakening feedbacks that would drive long-run sea level much higher than this, and could even trigger a runaway greenhouse effect, where emissions from fossil-fuel burning are triggering so much natural release of greenhouse gas from warming reservoirs like the permafrost that there is nothing emissions-reductions can do to stop the process.

I can't say a runaway effect won't happen, Schellnhuber says. Nobody has proved it can't.

There is utter silence in the room. I wonder what the chairman of Shell is thinking, sitting in the front row with the Prince.

Lord Nick Stern summarises the economics. Decarbonisation of the world economy is do-able, he says, at a cost of around 2% of gross global product. He had thought this figure would be 1% of GGP at the time of his Stern Review in 2006.

The UK special envoy on climate, John Ashton, a senior diplomat at the Foreign Office, covers the politics. It is quite clear that the goal now has to be zero carbon in energy, not 60% or even 80% cuts, he says. We need to build a zero emissions energy infrastructure at a time of resource crunch. We have to go zero in energy because of inevitable emissions in food production.

All the eminent speakers portray climate change to the assembled business representatives as a full-blown slow-burn end-of-civilisation drama. It has always been so, of course, but the days are long gone when only a few environment groups said so.

The business executives gather on the lawn for a buffet lunch. I always wonder, on occasions like this, what goes through the heads of people who have heard the full bad news on climate change for the first time.

The mood seems no different from any English social gathering on a lovely summer's day. We are British, for the most part, of course. A stiff upper lip is essential at all times.

An equerry asks me if I would like to meet the Prince.

I wait in a group of four for him to work his way round to us.

The equerry introduces us.

Ah, says Prince Charles, looking at me. You're the solar man. Can't you do something about those ghastly blue panels? Isn't there a way of making them more aesthetic?

The Prince of Wales is well known for his architectural tastes.

Indeed there is, Your Royal Highness, I say. We take solar cells and make them into dark roof tiles that look much like large slates. And in fact, you will be pleased to hear that we manufacture them in Wales.

Chapter 7

When the dancing stops

The dancing stopped on 15 September 2008. In the early hours of Day One of the financial crash, Lehman went bankrupt. Before the day was out Merrill Lynch had folded into Bank of America, and AIG – one of the world's biggest insurers – had begged the Federal Reserve to keep it afloat by paying its bills for failed insurance of credit.

The Fed may have let Lehman go under, but AIG had fully a trillion dollars on its balance sheet. The insurer was deemed too big to fail. On Day Two, the Fed loaned it $85 billion.

The money markets froze. Central banks pumped $200 billion into the global system. Former Financial Services Authority boss Howard Davies, a champion of the light touch regulation that had done so much to stoke this disaster, now suggested regulators faced just two choices: nationalise the banks or allow complete meltdown. Later forensic investigations showed that the entire banking system came within hours of complete collapse at the outset of the crash.

On Day Four, ratings agency Standard and Poor's warned that the worst was yet to come. This outbreak of forward-sightedness came from one of the agencies that had given triple-A ratings to so many of the mortgage-backed securities that were now turning toxic.

On Day Five, the US government stepped in to ring-fence toxic securities in a so-called 'bad bank'. This slowed the collapse of confidence in the plunging stock markets, but only temporarily.

Nick Leeson, the rogue trader incarcerated for the massive fraud that had brought down Barings in 1995, wondered in an op-ed in the *Guardian* (19 September 2008) who would go to jail this time round. Institutions and politicians were by now outbidding each other in their calls for recrimination. The FBI was combing the books at a handful of the biggest institutions already.

The answer to Leeson's question would end up being 'hardly anyone'. I shall return to that point.

In a rare confession, one sub-prime mortgage lender said he should be regarded as little better than a mid-rank drug dealer.[1] By that analogy the investment bankers – those who dreamt up the complex derivatives that had allowed so much lending on mortgages that were unpayable in a $40 oil world, much less a $140 oil world – would be drug barons. There were no confessions from them. Staggeringly, within just weeks of the crash investment bankers were earmarking huge bonus pools for their star 'risk takers'.

On Day 12 Washington Mutual filed for bankruptcy. It was the biggest US bank failure ever. Hedge funds were by now in deep trouble, with one estimate suggesting a third of all 10,000 heading for oblivion.

On Day 15 panic gripped the world markets again as Congress rejected US Treasury Secretary Hank Paulson's bank bailout plan. The bank failures continued. Citigroup bought Wachovia. The UK government nationalised Bradford and Bingley. The Irish government forestalled a run on Ireland's banks by guaranteeing all retail deposits.

Fear was now everywhere. The super-rich were draining the global gold supply. Gold refineries could not produce enough gold bars for people to shift into their own vaults.

On Day 17 the Senate voted the Paulson Plan through, but the stampede still continued. In the UK, cash flowed out of banks and into National Savings. Queues developed outside ATS Bullion's low-key premises next door to London's Savoy Hotel.

On Day 18 the short-term corporate finance market seized up, having shrunk by $95 billion in a week. This market, known as the commercial paper market, is where companies go to raise working capital for their goods and services. Even blue-chip companies like GE and AT&T were now finding it difficult to raise money.

On Day 21 Germany guaranteed all private savings accounts. Chancellor Merkel had criticised Ireland for doing this just the day before. Now it was every man for himself.

The great crash of 2008 was playing out faster than the great crash of 1929. The stock market crash then spanned three terrible days, and subsequently took three more years to hit rock bottom, by which time shares were 89% down in value. The 2008 crisis was playing out seemingly in a major hit a day.

On Day 24 the UK government part nationalised seven British banks, injecting £400 billion into them. On Day 25 the US followed suit, clearly fearing its $700 billion injection to buy mortgage debt would not be enough.

Exactly a year after the Dow Jones reached its highest ever point, it had lost a third of its value.

The extraordinary measures taken by the US and UK governments did not prove enough to prop up confidence. Day 26, 10 October 2008, became known as Black Friday. The FTSE fell 8.9% over the day, down 21% over a week that wiped £250bn off the value of UK companies. It stood below 4,000 for the first time in five years. Every European market lost at least 20%. The Dow Jones Industrial Average suffered its worst weekly loss ever, 18.2%, including the October 1929 crash. The Nikkei fell 23% over the week. Russia and Indonesia closed their stock markets completely.

The head of the IMF warned that the world financial system was teetering on the brink of systemic meltdown.

By now, conservative politicians were talking like socialists. Chancellor Merkel railed about the need to 'redirect the markets so they serve the people, not ruin them'.

The crash was unfolding in lurches of panic and response. On Day 28, 12 October, the UK government agreed to inject £37 billion into the British banks deemed too big to fail. The next day, Germany pumped €500 billion into its banking system and France €360 billion. The day after that, the US reluctantly bought minority stakes in its banks too.

Outrage was building all the while. A flood of lawsuits was backing up in the USA. Criminal investigations were under way in at least 15 companies. Stories emerged alleging sleaze between ratings agencies and banks. It was the triple-A ratings given by agencies like Moody's that had allowed the packaging of millions of dodgy mortgage loans into must-have bonds. Pension funds, seduced by these ratings, now held billions in structured products that were toxic. It seemed hardly surprising then to read about all the weekend getaways organised by investment bankers for ratings agency officials.[2]

A top hedge-fund manager fanned the flames. Andrew Lahde of Lahde Capital was thought to have made one of the biggest percentage profits ever by betting against the housing boom continuing. Lifting a metaphorical finger to the 'idiots' at the top of banks 'stupid enough' to take the other side of his bets, he announced he was now shutting his fund down to 'spend time with his money'.[3]

In the face of all this, it increasingly seemed as though the banks were oblivious, hide-bound in a culture encased in Teflon. Wall Street banks were still intent on huge bonuses for 2008: a total pool of $70 billion, fully 10% of the US bailout to date.

And still people said there was worse to come. One of these was Nouriel Roubini, one of the few economists who had blown the whistle ahead of the credit crunch.

On Day 43, 28 October 2008, the Bank of England announced that the financial institutions' losses in the crash to date totalled $2.8 trillion.

***

The drama in the capital markets, and everything it flagged about the dangers of ignoring or overlooking systemic risk, served to embolden the British companies that shared concerns about peak oil. We decided that if the UK government wouldn't join us in a risk assessment, then we'd do it by ourselves. In March 2008, representatives from the companies met in the boardroom at Solarcentury to discuss how best to proceed. At the time, not all of us agreed early peak oil was necessarily a high-risk threat. But what we all agreed was that even if the threat proved to be very low risk, it was certainly high consequence and therefore worth careful study.

We left that meeting having decided to complete a risk analysis by year end, and to name ourselves the UK Industry Taskforce on Peak Oil and Energy Security. The membership spanned a fair spectrum of industry: Virgin, SSE, Arup, Stagecoach, Yahoo, First Group, Foster and Partners, and Solarcentury.

By October we had completed our report, and everyone involved no longer thought of the issue as low risk. Our study had led us to the view that a clear and present danger lurked in the oil sector. We had come to the collective belief that peak oil could be expected sooner rather than later, meaning there was a major risk of global demand exceeding global supply, and a whole barrel of problems, for virtually every business sector.

In October 2008, the production figures of all five major international oil companies had been falling for five consecutive quarters, and there was clear scope for the national oil companies – the largest oil companies in the world, controlling some 80% of global production – at some stage to follow suit. Old oilfields and provinces were showing then that local and regional oil production can descend very fast beyond peak production, even where the best enhanced-oil-recovery techniques were applied. Meanwhile, industry was discovering fewer and fewer giant fields, notwithstanding the hype when they did discover one. Oil prices had been rising for four years, and as a consequence they had plenty of cash with which to explore, so that could not be the issue. Even when they did make big discoveries, we noted the long lead times before oil could be delivered to market: an average of six to seven years and sometimes in excess of ten years in the case of particularly large fields.

We hired Chris Skrebowski, Editor of the *Petroleum Review*, to update his plotting of all the major reported oil discoveries, to add the vast majority of the smaller discoveries, and to chart their due-dates for coming online. From this we subtracted the depletion rate in existing fields. That process left us with the rather alarming observation that net global oil flows slowed in the period 2011–13 and dropped thereafter. If demand kept on rising, it would soon outstrip supply. We could not understand why this was not galvanising governments and wider industry into a response.

On top of these geological concerns could be added geopolitical concerns, among which restriction of supply by decree of supplier governments featured high on the list. Beyond Russia's tendency to threaten withholding oil and gas exports, one particularly worrying issue involved supply from the world's number one producer. In April 2008, King Abdullah of Saudi Arabia had dropped something of a bombshell. In an interview he had said: 'I keep no secret from you that when there were some new finds, I told them, "no, leave it in the ground, with grace from god, our children need it".'[4]

The implications of that action for oil-importing industrial countries, were it to be turned into practice at any scale, hardly needed spelling out. Saudi Arabia holds spare capacity: a reserve of production capacity against which to lift production, if needed, in an effort to hold down the global oil price. The reserve is only thought to be around two million barrels a day today: little over 2% of the current world production of around 90 million barrels a day.

Fortunately for the rest of the world, the Saudi government has not yet actioned the King's thought.

The first UK Industry Taskforce report was ready for publication by October 2008. Our view, we knew, was a minority one. Most of the oil industry argued, then as now, that it would be able to meet global demand for oil far into the future. That view was accepted either explicitly or implicitly by almost every government, corporation and household on the planet.

Not everyone in the oil industry was happy with such a comfortable narrative, however, then as now. A former chairman of Shell, Ron Oxburgh, wrote the foreword to our report: 'In the past these views might have been regarded as heretical', he observed. 'But they are not and their warnings are to be heeded.'[5]

London Stock Exchange, 29 October 2008

I sit on the stage with four CEOs and chairmen, who are speaking for companies measuring their collective value in the multiple billions. Backing us up are all the PR resources such companies can muster when they try. Their press officers usually have the ability to persuade journalists to attend, and even sometimes influence them to write what the company wants to see. In case not, videographers are on hand to record the words of wisdom from the executives and post them on websites.

We are there to warn about a grave risk to the global and national economies.

It is Day 44 of the great financial crash. We are hoping that the press might have developed a taste for some news of a different kind, even if it is bad.

But the silent message from all the missing journalists at the press conference seems clear.

We can only deal with one global risk at a time.

By November 2008, world trade was seizing up. In America, car giants including GM were heading for the inconceivable: bankruptcy. In China, two-thirds of the recently minted billionaires had been wiped out. In the UK, housebuilding had halved. Government bailout funds now exceeded four trillion dollars in the US alone.

In the banking sector, write-offs exceeded a trillion dollars. British banks had become what *FT* columnist Martin Wolf saw as undercapitalised hedge funds with liabilities big enough to destroy the solvency of the British state. RBS, HBOS and Lloyd's were all ceding equity to the government to inject cash onto their near-death balance sheets. Barclays was desperate not to. Instead, they sent their top guns to the Gulf to sell equity to Gulf states. Investors were furious, fearing they would sell up to a third of the bank to Qatar and Abu Dhabi just to keep leeway for bonuses that British government shareholders would never allow them.

It emerged that four British banks were accruing billions in bonuses. RBS, perhaps the most disgraced of them all, had £1.7 billion lined up for the first six months of the year for a division that lost £5.7 billion in that period. This was nearly 10% of their £20bn bailout. Bank bosses professed that their star risk-takers needed this incentive to perform. Secretary of State for Business Vince Cable told them, and the press, that they were making monkeys of the government.

On 28 November, RBS was nationalised.

Singapore, 13 November 2009

Barclays Asia is wining and dining its Asian clients. A seminar has been organised for them at which the exciting events of our times will be mulled over by invited experts. I am in the role of what they term a friendly critic: the green-but-suit-wearing business guy who says challenging things without frothing at the mouth.

They may see a polite exterior, but inside I nurse feelings that are far from friendly.

I play golf in the pro-am of the golf tournament they are sponsoring at the time. I have a past in golf: a misspent youth hitting golf balls when I would have been better off studying. I eat the exotic Asian foods on offer at the sumptuous champagne receptions. I watch acrobats perform mid-air contortions sixty feet up satin ropes. The absence of safety nets seems somehow unsurprising. Gorgeous Asian women in cocktail dresses drift around. I find they don't seem to know much about the markets.

I stare at an ice statue, as tall as a man, slowly melting. It seems to encapsulate the guilt I feel at being here. All this excess, in a time investment bankers are scrambling under the skirts of governments all around the world, commanding vast bailouts involving the taxes of people who will never come close to luxury like this. Many of those people are in the process of losing their jobs, their homes, their hopes, directly as a result of the irresponsibility of bankers like many of the men and women whose company I am keeping.

Barclays' own spinmeisters are in the Gulf right now, trying to conceal their desperation while pitching for the fresh capital they need to stay out of the clutches of the British government.

I am invited to a small dinner for the Barclays senior management and their guest speakers at the British High Commission. I discuss the golf with Bob Diamond, President of Barclays, head of their investment banking.

Henry Kissinger is the guest speaker. I listen to the growly voiced old man dispense his hawkish view of the world, sitting but ten feet from me. I try to picture him with Richard Nixon, plotting the secret bombing of Cambodia.

He takes a few questions. I ask him what he sees in the crystal ball when it comes to oil supply. I am not surprised by his answer.

If demand for oil continues to grow, Kissinger says, a contest over access is inevitable.

A contest. So that's what the hawks call a war these days.

Lee Kuan Yew, the Singaporean premier, listens to this. He prefers to talk about climate change. A serious matter, he argues. There will be more coal. Therefore there must be carbon capture and storage. I am struck by how similar he sounds to a Big Energy chief executive.

The seminar starts next day. Bob Diamond kicks things off. So far he has escaped the mess in the markets, so the newspapers tell us.

Things are getting better by the day, Diamond says. But it will take time.

Jim Rogers, co-founder with George Soros of the legendary Quantum Fund, is the keynote speaker. He is a man with a point of view, offered enthusiastically.

The twentieth century was America's. The twenty-first will be China's. The US goes deeper into debt by a trillion dollars every 15 months. They will keep on printing money until the printing presses run out of trees. America is out of control. Asia is where the major creditors are. New York will end as a financial centre. So will London.

As for advice to the clients: forget bonds. Stocks have been flat or down for the longest time. It is commodities that can yield you money in these times. If you know what you are doing.

> What a game, I think to myself. A rising oil price drives many commodities markets up too: just look at the food prices around the world in recent times. Invest on the up. Then exit before any 'correction' in the market. And so you make a ton of money in a rigged casino, on the back of otherwise economically ruinous oil prices.
>
> And of course, there are periods in the casino when speculation forces up food prices. Periods when the gamblers, to all intents and purposes, starve the poor.
>
> How functional. What a great way to build a resilient society. To ensure social justice. To foster common security. To deprive the latter-day Henry Kissingers and James Schlesingers of their chance to throw us all into the meat grinder of a Third World War.
>
> Of course, I reflect, the Rogers view on commodities might just explain why I have been invited here by Barclays.

At the time of writing, Barclays is under investigation by the UK's Serious Fraud Office for their 2008 Gulf deal. They stand accused of lending the Qataris the money to make the equity purchase, an illegal practice. Bob Diamond faces trial separately for his role in the Libor scandal. All this squalor is yet to come in our history.

Entering December 2008, the drop in world trade and economic activity generally had been so swift that the oil price had fallen back to $50 for the first time since 2005. Ministers in the Gulf states said the world would be sure to face a supply shock if the price stayed below $70 for long: there wouldn't be enough cash for exploration and production.

The news on climate change, meanwhile, was little noticed. The Chinese government urged the West to move faster with emissions limitations. The British signed into law their national targets, the first nation so to do: 80% cuts in greenhouse-gas emissions by 2050, with interim targets to be met along the way.

Young people driven to despair by the news on climate change were not assuaged by this. A saboteur crashed a turbine in the coal-fired Kingsnorth power station for four hours, forcibly cutting British emissions by 2%. Protesters occupied the runway at Stansted airport, briefly shutting it down.

Poznań, Poland, December 2008

The community of nations has been talking for more than 18 years now about how to stop humanity's remorseless effort to cook its own home.

I take the train to Poland for the annual climate summit, a prospect that sounds like a recipe for slow-travel hell, but in fact is both easy and productive. You take the afternoon Eurostar to Brussels, the evening express to Cologne, the night train to Poland, disembark after eight hours' sleep just in time for breakfast, with a massive reading backlog dismantled along the way.

At the talks, thousands of delegates throng in cavernous halls, trying to find out what is going on behind the closed doors of the intergovernmental meetings where most of the serious negotiating is done. Hope is in the air. The EU's 27 leaders have just agreed a '20 20 20' deal in Brussels: 20% cuts in greenhouse-gas emissions union-wide and 20% renewables in the energy mix, both by 2020.

One of my missions is an effort to raise the peak-oil issue. Most of the 9,000-plus attendees – diplomats, lobbyists and journalists – will have little idea how strong the evidence is that a global energy crisis looms, and that it could greatly help their climate policymaking, albeit amid very hard times.

Some of that evidence is aired in Poznań by the International Energy Agency at an open meeting on its recently completed *World Energy Outlook 2008*. Between the lines of the IEA's latest weighty annual lies an early warning of a premature peak in global oil production. I say 'between the lines', because the IEA is a somewhat inconsistent organisation. Set up by developed governments essentially to promote fossil fuels, it has to wrestle with considerable internal tensions when warning both of fossil-fuel depletion and the environmental impacts of fossil-fuel burning. These tensions are often discernible in the wording of the agency's committee-written reports, and in public presentations by its officials.

This year, for the first time, the IEA has conducted an oilfield-by-oilfield study of the world's existing oil reserves. It shows that the fields currently in production are depleting alarmingly fast. The average depletion rate of 580 of the world's largest fields, all past their peak of production, is fully

6.7% per annum. They had not been expecting this when they started the exercise.

In a side seminar, IEA Executive Director Nobuo Tanaka shows a slide illustrating the situation. It is, he says, his most important diagram. It shows crude oil production from all the world's existing fields climbing unevenly from just below 60 million barrels a day in 1990 to a peak – more exactly a brief plateau – of just over 70 million barrels a day between 2005 and 2008. In 2009, however, crude production begins a steep descent, falling steadily all the way below 30 million barrels a day by 2030. The depletion factor charted by his team, as I see it, could better be called a fast-emptying factor.

This is indeed alarming, Tanaka says. The more so because, even with demand for oil being destroyed fast by recession in the West, the rate of demand growth – led by China and India – is such that the world will need to be producing at least 106 million barrels a day by 2030.

Can that be done? he asks.

I study him carefully as he speaks. His body language is revealing.

Yes, he says slowly. But only if massive investment is thrown at the challenge, especially by the OPEC nations. Global production today totals 82.3 million barrels a day if we subtract biofuels and add to existing crude production the 1.6 million barrels a day of 'unconventional' oil squeezed from the tar sands and 10.5 million barrels a day of oil produced during gas-field operations. To reach production of 106 million barrels a day by 2030, therefore, would require oil-from-gas to expand almost to 20 million barrels a day, unconventional production to expand almost nine million barrels a day, and on top of that more than 45 million barrels a day of crude oil capacity yet to be developed and yet to be found. All this adds up to 64 million barrels a day of totally new production capacity needed onstream within 22 years.

That, says the IEA's Executive Director, pausing for effect, is fully six times the production of Saudi Arabia today.

I imagine I can detect a desire in Tanaka to say more about his thoughts on the likelihood of this. But of course, in his position, he can't.

Here is the bottom line. At oil prices below around $70 a barrel, producing oil becomes uneconomic in many settings today. With the oil price where it currently languishes, at less than $50 a barrel – in a market where pricing has become completely disconnected from 'fundamentals' by the volume of paper trading – oil development and exploration projects are being cancelled around the world on a daily basis. How on earth is the industry going to bring on six new Saudi Arabia's from this kind of dead-in-the-water start?

That is before you even consider all the other concerns harboured by the UK Industry Taskforce.

Tanaka closes by saying that the world needs a 'clean energy new deal', as the IEA is calling it. Insurance must be taken out, via clean energy, in case the oil industry fails to meet projected demand. The perils of climate change require such a course of action anyway. So too does the rebuilding of economies made necessary by the financial crisis. It all makes sense in a win-win-win sort of way.

I ask Tanaka whether he knows of the recent study by a group of eight UK companies, the UK Industry Taskforce on Peak Oil and Energy Security. These companies, including my own, have conducted a business-risk assessment of the likelihood of the 'six Saudi Arabias' being found. Our conclusion is that it is unlikely that the oil industry will close the widening gap between depletion and demand within a few years.

There is a risk, as you say, of a constraint on the supply side, Tanaka replies cautiously. We hope the climate-change issue will drive the world to take proactive action, he says. It's a choice: peak oil or you yourself (meaning the community of nations) will drive energy efficiency and alternatives.

Tanaka hadn't mentioned the words 'peak oil' once in his presentation. Only now, in discussion, did the seemingly taboo term emerge.

Afterwards, an IEA official comes up to me.

There's a real risk that this thing is going to collapse, he says. He means the operating model for the world's energy markets. Where financial markets can go today, in other words, so can energy markets tomorrow.

Perhaps 100 of the 9,000 delegates in Poznań attend Nobuo Tanaka's presentation.

Chapter 8

This house believes

The stock markets of the world suffered their worst year on record in 2008. Entering 2009, $14 trillion had been wiped off the value of shares in a single year. The FTSE had lost almost a third of its value. With suicides on the rise in the financial sector, mental health professionals were reporting a surge in business. At least twenty-five companies were under federal investigation in America. Bernie Madoff, long touted as a Wall Street investment genius, admitted in December 2008 that his $17bn fund was 'one big lie', a giant ponzi scheme. In January, it emerged that Madoff feeder funds had been backed with billions of dollars of loans from HSBC, RBS and other banks. But ominously, none of the main Wall Street investment banks had invested in him. Could it be, commentators asked, that they knew he was a fraudster and kept quiet for fear of stopping the dancing?[1]

And still it got worse. On 16 January 2009, the crisis entered a new phase of panic. Prime Minister Gordon Brown, staring at the prospect of a bankrupt nation, ordered the British banks to come clean about the extent of their bad assets. He had been asking for a year, it turned out. The Bank of England announced it would be buying government bonds, or gilts as they are known, from banks on a massive scale — so-called quantitative easing — as a way to inject cash into the failing system.

And not a single British bank boss had yet apologised for the trauma they were inflicting on society.

As for the oil price, it had fallen to $35. At those levels, only Exxon and Total would be able to finance their investment programmes from cash flow.

So, the early peakists wondered, who was going to be doing the exploration that would be needed to find the six new Saudi Arabias that the IEA was talking about?

As for the prospect of cheap gas, Russia had joined with Qatar and Iran to form a gas-producers' equivalent of OPEC. Hosting a meeting in Moscow in December 2008, President Putin told the West that the era of cheap gas was over. In January, he turned the pipelines through the Ukraine off again. Tens of thousands of Europeans shivered once more in unheated homes, hostages to the Kremlin's recurrent willingness to use oil and gas as an offensive weapon.

Oslo, Norway, 6 February 2009

The snow is metres deep outside. The log fires are yards across. The food and wine are world class. And the company isn't at all bad either.

Every year a Norwegian foundation invites a hundred oil and gas industry bosses to a mountain retreat above Oslo to discuss the state of the global industry. I am an invitee this year: the just-about-acceptable face of their critics.

I meet the charismatic, larger-than-life CEO of Total, Christophe de Margerie, for the first time. We are together on a discussion panel of four one afternoon: that's how the Norwegians run the event. I want to explore his discordance with the rest of his industry on peak oil. He is the only oil industry boss who admits to any possibility of a problem. But he doesn't want to play ball. He wants to dive into the reasons why someone like me can have such a low opinion of nuclear power. Surely this has to be a vital part of the future, he argues. I mean, just look at France, where 80% of the electricity is nuclear.

You should be worried, not pleased, I counter. And I try to talk about peak oil again.

One night, after dinner, I am in a mood to stay up late. So too is Bob Dudley, it seems. One by one the brandy drinkers head off to bed, until it is just me and Bob, chatting about life and times by the dying fire.

I have known this quietly spoken man for a while now. He ran BP's solar arm before he headed off to Russia to run TNK-BP. There he endured a campaign of persecution and intimidation by both the authorities and the oligarch owners of TNK, the Russian half of BP's joint venture. BP's office in Moscow was twice raided by armed police. Dudley came under criminal investigation on tax charges. All this culminated in his being refused a re-entry visa in July 2012.

TNK-BP has been the vehicle delivering almost all BP's reserve growth in recent years. Now it looks increasingly as though BP will lose the whole enterprise to Putin's creeping programme of back-door resource nationalisation.

I look at Bob Dudley in the firelight. I feel a quiet admiration for him, despite myself and all my issues with BP. This man tried to run a business honestly, so far as I can see, in what many describe as a mafia state. Suspicions abound that the horrific assassination of Alexander Litvinenko in London, by radiation poisoning in November 2006, was perpetrated by agents of the Kremlin. Dudley must have been brave indeed to take on the TNK-BP job.

So Bob, I say. Was there ever a time while you were in Russia when you feared for your life?

Dudley looks at me with clear eyes and a faint smile.

Yes, he says after a moment.

And he tells me about it.

As he tells his story, I picture Tony Hayward, Dudley's boss, in BP's boardroom back in November 2004, assuring me about the reliability of Russian oil supply because of his cosiness with the Kremlin.

By early 2009, post-mortems into how the financial crash had happened were well under way. Bankers tried to pass much of the blame to credit rating agencies, even as accounts were emerging of risk officers in banks being ignored at best, fired at worst.[2] RBS joined those facing the prospect of criminal investigation, after non-executive directors complained of intimidation and threats.[3] A spotlight fell on the business schools, wherein many of the architects of the disaster had been trained in both high finance and, so it was alleged, the special kind of arrogance needed to perpetuate the myth that risk had been excised from complex derivatives.

President Obama, newly inaugurated, found himself a man drinking deep of a poisoned chalice. In February 2009, he signed a $787 billion stimulus package into law. He described his hopes for it to the press on a rooftop solar installation in Denver. The creation of green jobs, he said, had to be a major factor in the fight to repair the US and global economies. In London, similar sentiments were aired by the Brown government.

In March, the Bank of England began their quantitative easing strategy. The crisis still hung in the balance.

Queen Elizabeth II Congress Centre, London, 31 March 2009

A debate at the Seventh Petroleum Geology Conference. The motion: 'This House Believes Peak Oil Is No Longer a Concern'. A former chief geologist at BP, David Jenkins, argues for it. I argue against. The audience, comprising several hundred practising geologists and geophysicists in and around the oil industry and related academia, will vote for or against at the end.

I have said yes to this invitation with no hope in my heart that the debate is winnable. I am not looking forward to it. I know my opponent quite well. He is one of the industry's most respected geologists, a suave character with a sharp mind and a ready smile. He and I have debated climate change in private and public for years. I like him, and not just because he was one of the many BP executives who changed their minds about climate change in the 1990s. When BP became the first oil company to break away from the carbon club in 1997, professing that climate change is indeed a problem for society, then-CEO John Browne announced the change in policy in a speech at Stanford University. David Jenkins wrote large sections of that speech.

I arrive in time to circulate before the debate. On the rare occasions I venture on this kind of excursion into my past, it always amazes me how fast people age. Lads I had known as students are now middle-aged men.

How can that be? Am I supposed to treat them any differently to when they were students?

I certainly find myself unable to think of them differently. Nobody you have been drunk with in a student bar can expect to be taken too seriously three decades later, no matter how many oil company departments they have run in the interim.

I greet old faces with as big a smile as I can muster. I have to soak up a lot of banter on these occasions, but none of it is ever particularly malicious.

The oil industry doesn't have the general mean-mindedness of the coal industry.

Jenkins is not someone I knew as a student. I marvel at how much more default respect my brain awards him for that. I am sure his delivery will be as immaculate as his appearance.

He speaks first, and his opening surprises me a little.

The days when the OECD could expect to grow its own oil supply are over, he says, and will never return. The world needs to become accustomed to expensive energy. If the present concerns about carbon dioxide persist, the requirement to decarbonise carbon-based fuels will mean a further significant increase in costs. In fact, rather than the supply peak over which commentators have fretted, high energy prices could even engender a demand peak for oil, something which the environmental movement would no doubt applaud.

He looks across at me from the podium and smiles.

But the fact is, he continues, that there is no near-term resource peak for oil.

We only have an hour for the whole debate: eight minutes each to state a position, with the rest being a question-and-answer session with the voting audience. Jenkins majors on peak demand. It is the key BP argument when combating the idea of premature peak oil.

Moving away from the internal combustion engine to an electric drive train, associated with greatly increased production of biofuels, he says, will soon be technically feasible and could very plausibly produce a scenario that eliminates further growth in global demand for oil.

Zero growth demand for oil. I scan the audience for Chinese and Indian faces to see how that argument might play with them. This is an international conference, but almost all the faces are Caucasian.

Combine this with the production capacity available from the global resource base and a future world of expensive carbon-free energy, Jenkins concludes, then peak oil clearly becomes an artefact of a prior economic world. In this somewhat bleak future it will definitely not be one of our many concerns.

The wall of applause as he finishes sounds highly appreciative to me. I think to myself, not for the first time, that it must be very comforting for industry practitioners to view themselves as part of a solution, not a problem.

The chair of the debate, Channel 4's Science Correspondent Julian Rush, invites me to respond. Julian and I have drunk a lot of beer together at climate summits. He raises his eyebrows to me in a manner that conveys 'rather you than me, mate'.

At the podium, I survey the faces. There are lots of smiles out there. I take them to mean 'how the hell are you going to follow that?'

I explain that those of us who worry about peak oil fear that the oil industry has lapsed into a culture of over-exuberance about both the remaining oil reserves and prospects of resources yet to be turned into reserves, and about the industry's ability to deliver capacity to the market even if enough resources exist. Our main argument is that new capacity flows coming onstream from discoveries made by the oil industry over the past decade don't compensate for depletion. Hence projections of demand cannot be met a few years hence. This problem will be compounded by other issues, including the accelerating depletion of the many old oilfields that prop up much of global oil production today, the probable exaggeration by OPEC countries of their reserves, and the failure of the 'price-mechanism' assumption that higher prices will lead to increased exploration and expanding discoveries.

In 2008, I observe, the International Energy Agency conducted an oilfield-by-oilfield study of the world's existing oil reserves for the first time. I quickly recount the same story that the IEA told the climate negotiatiors in Poznań. Surely they know this, I tell myself. But maybe not all of them.

The oil industry is not discovering giant oilfields at anything like the rate it did in the 1960s – the peak decade for discoveries. This is the case even with much better equipment for exploration today, and even after four years of rising oil prices from 2004 into 2008, when exploration was not hampered by lack of funds for investment.

In addition, the oil industry has profound infrastructure problems, and major issues with underskilling and underinvestment. Many drilling rigs, pipelines, tankers and refineries were built more than 30 years ago, and according to some insider experts the physical state of the global oil infrastructure is a major problem even at current rates of oil production, much less the significantly higher levels anticipated in the future. The average age of personnel in the oil industry is 49, with an average retirement age of 55 – little less than a human-resources time bomb.

I survey the audience. There is an awful lot of grey hair out there.

To add to the challenges, the industry's overall exploration budget has actually fallen in real terms in recent years.

The UK Industry Taskforce on Peak Oil and Energy Security, for one, fears that these issues will synergise to compound the peak-oil crisis, gravely impairing society's collective ability to respond.

In conclusion, this debate is all about the risk of a mighty global industry having its asset assessment systemically overstated, due to an endemic culture of over-optimism, with potentially ruinous economic implications.

I pause to let that sentence hang in the air for a second or two.

Now that couldn't possibly happen, could it?

This none too subtle allusion to the disaster playing out in the financial sector elicits a polite laugh from the audience.

Into the debate we go.[4] It whistles by. So many issues. So much knowledge under one roof. So many men who seem to like the sound of their own voices.

It is time to vote. The audience has been furnished with coloured cards to wave.

Those in favour of the motion – peak oil is no longer a concern – please raise your cards, Julian Rush says.

I am expecting to see a forest of them, but only around a third put their arms up.

I am amazed. There must be a lot of abstentions, I think.

Those against the motion, please raise your cards.

Around two-thirds. Few abstentions, if any.

It takes a moment to sink in. The industry's front-line practitioners do not believe their own leadership's comforting narrative on peak oil.

I do my round of handshakes, and trot out the usual promises to make a better job of keeping in touch with faces from my past.

I wander out into Whitehall with Julian Rush. In the tea room under the Methodist Central Hall we reflect on our shared experience. He is as amazed as I am.

I do not delude myself that my rhetoric or arguments made any difference at all, I tell him. People of knowledge come to beliefs, and those beliefs are invariably difficult to shift once arrived at. What we were dealing with was an opinion already held by a majority.

Why on earth don't they speak out?

Chapter 9

We will be blamed forever

City of London, I April 2009

Across London Bridge, towards the Bank of England, a forest of banners flutters in the breeze among the marching thousands. The slogans have a clear theme: capitalism isn't working, and the leaders of the G20 – in town for a summit on the financial crisis – should fix it fast.

I walk alone among strangers. They are of all shapes and sizes, a complete cross-section of society from retirees to schoolchildren. The accents I hear are mostly middle class. Notably absent are people in business attire, who instead line windows above the streets, looking out at the protesters. One man waves a fistful of banknotes. He triggers a barrage of whistles.

Near the Bank there is talk in the crowd of a police 'kettle', whatever that means. I join a throng slipping up a side street to Bishopsgate, where hundreds of students are due to erect what they are calling a 'climate camp' of tents across and along the closed road.

A TV journalist and her cameraman approach me. Eyeing my pin-stripe suit and the rolled up copy of the *Financial Times* I am carrying, she asks me if I am protesting or just watching.

Protesting, I say.

She asks if I would be willing to be interviewed by CNN. I say yes. Stand by then, she says. She speaks into a headset microphone, telling what I assume to be her producer that she has found a businessman who is a protester. I realise that the short aerial on the back of her crewman's camera must be capable of live relay. I didn't know they could do that.

She straightens, and asks me what somebody dressed as I am is doing here.

I presume I am live on TV, though she has never mentioned this.

I am protesting, I say, because modern capitalism is broken, and needs to be re-engineered root and branch. The G20 leaders have a window of opportunity to do that, and they must seize it.

Moving on to Bishopsgate, I watch as a technicolour tented protest town springs up like a military operation. I find a place to sit among the tents and read my papers. I have brought a stack of reading along, and a picnic lunch.

Some hours later the CNN lady reappears. She wants to do another live interview, she says.

This time, as we wait, she is in a conversation with her producer about what to film next.

But there aren't any riots, I hear her say.

Indeed, I think. I haven't seen a single violent act along miles of the City's protest-clogged streets. That must be pretty disappointing for both the media and perhaps many of the Robocop-lookalike policemen deployed in battalions in side streets around the demonstration.

At the tube station on the way home, the *Evening Standard*'s front page headline is a puzzle: 'Riot Police Battle Anarchy in City'.

The G20 summit was widely deemed a qualified success: 'the first bricks in a new world order', as an *FT* editorial put it. The leaders agreed to inject $1.1 trillion into the world economy, entirely for developing countries. Their communiqué – nine pages long, hammered out over two days – claimed that the $5 trillion of stimulus money already deployed, plus the new $1.1 trillion, would get global output back on track. The global banking system would be reformed, they pledged, with controls on hedge funds, better accounting standards, and tighter rules for ratings agencies. Tax havens not sharing information would be named and shamed. An obscure Basle-based banking network of central bankers and regulators would be rebadged as the Financial Stability Board, and told to work alongside the IMF on restoring order.

But the leaders failed to agree new and binding measures to purge the toxicity of banks' balance sheets. And many banks, it was clear, still did not know the full extent of their most toxic assets at this time.

What wasn't a success was the behaviour of the Metropolitan Police. As the days went by, video evidence emerged of unprovoked police assaults on students, police lies about a man shoved violently to the ground from behind who later died, and testimony that the police used agents provocateurs to try to incite protesters to violent acts.

Three weeks into the post-mortem, a former Scotland Yard commander wrote that the police in modern Britain were being trained as though peaceful public protest is illegitimate. Because of a crisis of leadership, he wrote, 'officers are trained to regard every situation, no matter how benign, as a threat situation. The lesson is that the public are your enemy.'[1]

**

By the time of the 2009 G20 summit the economic crisis had pushed the number of chronically hungry people in the world over a billion for the first time. Even in the US, one in ten citizens was receiving food stamps. Off Singapore, the largest fleet ever assembled stood idle: 735 empty tankers marooned by the receding tide of world trade. Comparisons of the first year of the current crisis and the first year of the Great Depression showed that the two tracked each other depressingly closely.

Many expected the recession to be good news for oil prices. 'Cheap oil forever', *Newsweek* announced excitedly in a front cover headline in April. 'Why prices will keep on falling – and falling.' Correspondent Ruchir Sharma was much struck by the fact that the previous bull market in oil, in 1979, had been followed by a bear market that lasted twenty years. 'If history is any guide, we're only at the beginning of another one', he concluded.

In fact, the price was soon pushing back up again, and by June passed $70. As recently as late April Goldman Sachs had been predicting $45 oil within three months because of plentiful inventories and weak demand. Now the bank reverted to a bullish forecast: $85 a barrel by year end.

BP's profits were falling sharply in the face of the low prices of early 2009. Tony Hayward announced predictably that the company would be cutting back on exploration. The OPEC view was equally predictable: exploration and development cutbacks would harm the prospects of supply meeting demand after the recession.

Oil will peak because of peak demand, not unavailability of supplies, Hayward asserted.

Hayward was cutting in other areas too. In May, the BP boss announced a further milestone in the recarbonisation of his company. 'I think solar is probably the most challenged of all of BP's alternative energy interests', he told a conference in California. 'It is not going to make the transition to be competitive with more conventional power, the gap is too big.' There would need to be a step change in technology, he said. [2] BP had been shutting solar factories around the world and no longer set a target for solar sales. Hayward was widely interpreted as positioning BP's non-biofuels renewables activities for the chop, not least by my informants in BP itself.

Hayward's assessment sat uncomfortably with reality in the renewables industries, which were continuing to grow fast. In 2008, both the United States and the European Union added more power capacity from renewables than from conventional sources, including gas, coal, oil and nuclear. At least 64 countries had policies to promote renewable power generation, and renewables policy targets had been set in at least 73 countries. The favourite policy involved so-called feed-in tariffs: premium-price payments funded by levies on overall bills, designed to reduce in regular increments as the volume sales stimulated pushed the prices of renewable energy down.

Given these growth rates, it was becoming increasingly clear to me that renewable energy was beginning to appear on the radar screen of energy-incumbency bosses as some kind of threat. In April 2009, EDF and Eon confirmed my fears by openly recommending to the British government that renewables be cut back in favour of nuclear.

As for Shell, they announced in April 2009 that they would be pulling out of even conducting research on solar and wind power. They may not have been responsible for the energy the world uses, but they surely now had views on what forms of energy were best for it.

Groningen, Netherlands, 17 June 2009

Shell and ExxonMobil are celebrating the 50th anniversary of the Groningen gas-field. It is the 20th biggest gas-field in the world, one of the biggest discoveries of the twentieth century. They operate it jointly and are proud of their achievements. How better to celebrate than with a one-day conference.[3]

Shell's Vice President for Global Business Development describes the company's latest scenario-planning work. He lays out what Shell calls the Scramble scenario: a free-for-all future wherein every player pursues their own perceived short-term interests, there is little collaborative effort to cut greenhouse-gas emissions, and concentrations of greenhouse gases end up at around 1,000 parts per million carbon dioxide equivalent in the atmosphere by 2100.

Then there is the Blueprints scenario. In that version of the future, a co-ordinated effort by governments ends up with us facing 660 parts per million carbon dioxide equivalent in the atmosphere.

As we all know, however, he says, a maximum safe amount of carbon dioxide in the atmosphere has been set by the EU and many others as around 450 parts per million, in order to keep the maximum temperature rise due to global warming at under 2°C.

The organisers have elected to use technology to test the opinions of the several hundred people present. They can vote anonymously and give instant quantitative feedback.

Which scenario most closely resembles the course we are following, Scramble or Blueprint, the audience is asked.

Some 68% of them opt for Scramble.

My turn to speak. If we are going to avoid going over the 450 ppm danger threshold, I say, we can only extract about a third of the remaining known reserves of coal, gas and oil, absent the increasingly unlikely prospect of industrial-scale carbon capture and storage. We need to go way beyond Blueprint.

And your company has just pulled out completely from the main mechanism for doing that.

Bernard Madoff was jailed for 150 years in June 2009. In an interview with the *FT* from his cell two years later he would profess that he got away with his fraud for so long because what he did was deliberately ignored by top bankers on Wall Street.

At the time, it was as though Madoff's sentencing was the sole necessary catharsis. On Wall Street and in the City of London bank reform was running into the ground. As the profits began rolling back into banks, bankers defended their bonus culture passionately. Goldman Sachs told its employees they could look forward to record bonuses at the end of the year.

In August, with the credit crunch two years old and banks still lending only a dribble to small and medium-sized companies, insolvencies of British companies reached a record level. The chairman of the FSA was moved to call bankers 'socially useless' in a speech calling for their bonuses to be taxed.[4] In September, a poll showed that savers were losing faith in banks, building societies and financial advisers. Another showed trust in business leaders generally at an all-time low.

The Governor of the Bank of England now professed that the fallout from the financial crisis would take a generation to work through. In November, fearing the G20 was abandoning its intent to reform, the Bank issued its most urgent warning yet. Executive Director for Financial Stability Andy Haldane said the banks must be reformed with measures as sweeping as those in the Great Depression, or they will 'game the state' over and over in a 'doom loop'.[5]

***

The Bank of England was not alone in the institutional whistleblowing. In June 2009, IEA boss Nobuo Tanaka warned of the potential for an oil supply crunch by 2014, should global GDP growth hit 5%. If it only grew 3%, he added, the world could expect a postponing of the supply crunch until after 2014.

In 2010 global GDP growth was 4.34%. In 2011 it was 2.73%.

In August, the British government published its own view of energy security risks. In a report commissioned by the Prime Minister, Malcolm Wicks, the former energy minister, concluded that 'there is no crisis'.[6] The Wicks Review mentioned peak oil only once in the 122 pages of the report.

The relevant passage concluded: 'Few authors advocating an imminent peak take account of factors such as the role of prices in stimulating exploration, investment, technological development and changes in consumer behaviour.'

The UK Industry Taskforce on Peak Oil and Energy Security report of 2008 had ignored none of these things. Prices do stimulate exploration but – we argued – not enough. We discussed the intervals between oil discoveries and bringing capacity to the market. We discussed investment and concluded that there had been dangerous shortfalls even when prices had been high. We discussed technological developments such as enhanced oil recovery and concluded that they tended only to slow depletion rates. We discussed changes in consumer behaviour and worried that they would not be sufficient, especially in India and China, to shrink global demand in parallel with supply.

The Taskforce had held two meetings with Department of Energy and Climate Change (DECC) officials during their preparation of the Wicks Review. Wicks had attended one himself. We had said in our 2008 report: 'The risks to UK society from peak oil are far greater than those that tend to occupy the government's risk-thinking, including terrorism', and this is what we had told the government in our meetings. The government had chosen to completely ignore us. Indeed, it had ignored not just our conclusions, but our very existence.

As I phoned round the Taskforce to see how the Wicks Review was being received, I have rarely heard business people more angry or perplexed.[7]

In September, Total CEO Christophe de Margerie joined the IEA in predicting an oil crunch as soon as 2014.[8]

It wasn't as if the UK government was ignoring whistleblowing from everyone except the oil industry. It was ignoring shrill warnings from within the industry itself. It was as though one of the investment banks had said in 2006: 'Beware, these derivatives will end up toxic and risk wrecking the global economy', and the White House had responded: 'Sorry, we'd rather pretend you hadn't said that.'

Of course, on the other side of the debate, the beguiling message of plenty appeared much more often than the warnings. In October, the message from the World Gas Conference was that recent shale gas discoveries had put off an energy crisis for years. Major discoveries of oil in deep water off Brazil were generating similar mantras about oil supply.

Many no doubt happily took this comforting news at face value. But in November, a senior IEA whistleblower spilt some serious beans to the *Guardian* newspaper's Terry MacAlister. The unnamed official said the agency had been deliberately underplaying a looming shortage for fear of triggering panic buying. He claimed the US had played an influential role in encouraging the agency to understate the rate of decline from existing oilfields while overplaying the chances of finding new reserves. The official questioned the prediction in the *World Energy Outlook* that production can be raised from its current level of 83 million barrels a day to 105 million barrels. As he put it: 'The IEA in 2005 was predicting oil supplies could rise as high as 120 million barrels a day by 2030 although it was forced to reduce this gradually to 116 million and then 105 million last year. The 120 million figure always was nonsense but even today's number is much higher than can be justified and the IEA knows this. Many inside the organisation believe that maintaining oil supplies at even 90 to 95 million barrels a day would be impossible but there are fears that panic could spread on the financial markets if the figures were brought down further. And the Americans fear the end of oil supremacy because it would threaten their power over access to oil resources.' A second senior IEA source, also talking to the *Guardian* on condition of anonymity, said a key rule at the organisation was that it was 'imperative not to anger the Americans. We have [already] entered the "peak oil" zone. I think that the situation is really bad.'[9]

***

In the USA, President Obama was endeavouring during the summer and fall of 2009 to make good his pre-election promise of action on climate change, aware that the annual climate summit at year end would be a particularly significant one. In Copenhagen, the future of the Kyoto Protocol was at stake. A climate law was passing through Congress. In June, the House of Representatives voted narrowly in favour of a bill committing the US to cutting carbon emissions by 17% from 2005 levels by 2020 and 83% by 2050. It was the first time US lawmakers had ever voted for action on climate.

The bill ran to 1,200 pages. It committed the US to establishing a national cap-and-trade system, wherein limits would be put on emissions and permits to emit could be traded. It also required power companies to produce 15% of their electricity from wind and solar.

The bill had yet to be passed by the Senate, and the incumbency reacted swiftly to ensure it wouldn't be. The American Petroleum Institute wrote to its member companies suggesting they 'move aggressively' to set up multiple 'energy citizen' rallies protesting against the potential legislation. The first of these rallies, attended by more than 3,000 oil industry supporters, took place in Houston. T-shirts in the crowd bore slogans like 'It's the job climate stupid' and 'Think job losses and $4 gas'.

ExxonMobil, meanwhile, had continued to fund climate denier groups despite their promises not to. There could be little doubt the constant disinformation was taking a toll. An October 2009 poll showed only 57% of Americans thought the atmosphere was warming, down from 77% two years before. This was the stuff of Orwell, but it was to get much worse.

In November, on the eve of the Copenhagen Summit, the denial campaign scored its biggest hit yet. An anonymous hacker stole thousands of e-mails between leading climate scientists, posted them online, and claimed they showed collusion between scientists to manipulate data and hype the climate-change threat.

Most of the 160 MB of data showed nothing untoward. But there was a bare minority of exchanges allowing doubt to be cast. It was all the perpetrators would need.

'Climate-gate' had broken at just the right time for the carbon club, and the worst possible time for world leaders booking their trips to Copenhagen.

Copenhagen, December 2009

I catch the ferry to Denmark for the climate summit that will decide the future of the climate negotiations beyond the Kyoto Protocol. Or, as many of the young people travelling on the boat with me evidently feel, decide the future, period.

It is easy to see why. If all governments do just what they have committed to most recently in terms of greenhouse-gas emission limitations, and no more, then they will place society on course for 3.5°C global warming and almost 800 parts per million CO_2 equivalent in the atmosphere by 2100, 650 ppm of it carbon dioxide itself. So concludes Germany's centre for climate research, the Potsdam Institute, in its latest report. Existing pledges of action would not halt emissions growth until 2040, 25 years after the 2015 target called for by the Intergovernmental Panel on Climate Change.

The Copenhagen Summit must move well beyond this. It must get the world onto a road to the deep cuts.

President Obama has said he will come for the endgame, the last few days, immediately opening up the space for hope. As things stand, the President is uncertain that he can manoeuvre his current best offer past an obstructionist Senate, and he will need to do a lot better than that 17% cut of 2005-level emissions by 2020 if he is to lead the way to a potential global deep-cuts regime. If he tries, he will face even more of a battle royal with senatorial climate-change deniers and foot-draggers. But he would also be able to hope that the signal he sends in the effort can tease out a global survival reflex.

I install myself in a flat in downtown Copenhagen for the two weeks of the summit. I will be blogging daily for the *Financial Times*.

I walk the corridors and talk to delegates. A week in, a sense of *déjà vu* descends on me. It feels like Kyoto in 1997. At that climate summit, the first week was characterised by a strange brew of seriousness of intent and diplomatic posturing. Brinkmanship was rife, and yet hardly any delegations wanted the talks to fail, then as now. The negotiations were scheduled to end on a Friday evening, and the protocol was finally gavelled through at approaching 10 the following morning.

It took the arrival of ministers in the second week of the Kyoto summit to unlock the big-ticket agreements. The outcome was a protocol capable of sending a big enough signal to the markets to mean very bad news for those who sought to defend the status quo at all costs. As I describe in the final pages of *The Carbon War*, the carbon club's lobbyists were distraught at the outcome.

This time there are two differences, one negative and one positive. In Kyoto, the negotiators were closer to a final deal. In Copenhagen, unresolved issues span all major areas: mitigation, adaptation, finance and legal matters. On the other hand, it will not be ministers who descend for the endgame this time, it will be heads of state, and from more than half of the 198 countries present. They will not want to leave without a meaningful deal.

The tens of thousands on the streets long for one. The marches are getting bigger by the day. Drums of protest drift recurrently across the city. A long column of police vans sits semi-permanently outside my flat. Helicopters are constantly overhead.

I write of my cautious optimism on the *Financial Times* website. So too does former BP CEO Lord Browne. I worry that I am sounding too similar to him.

The heads of state arrive. The conference centre goes into lockdown. The gun-toting and surly police begin to exercise gratuitous violence on the anguished students peacefully protesting. This is not the Denmark I had been expecting. I worry anew about policing in the modern world.

And the news drifting out on Twitter grows steadily worse. The Danish PM's office is in meltdown, promulgating chaos, without the diplomatic skills or experience to force a way forward, and having lost the trust of the developing countries. Tuvalu for one has given up. We will leave with a bitter taste in our mouth, the PM of the Pacific island nation tells a press conference. The most vulnerable have not been listened to.

It is clear whom he blames. It is amazing, he says, that the US has not considered humanity.

Hillary Clinton's speech, prior to Obama's arrival, is a pitiful thing to behold, for those who know the long history of the talks. The US is prepared to work with other countries towards a goal of mobilising $100bn a year to address the needs of developing countries, she announces, trying to cast this as a breakthrough.

This is all they can manage. Governments have spent $10.8 trillion on the bank bailout so far: an average of $10,000 for every one of the billion or so people in the richest countries. $9.8 trillion has been spent by governments in the rich nations. China and other emerging nations have spent $1.6 trillion.

Gordon Brown judges the stakes right. We will be blamed forever, he tells delegates, because human survival is at stake. In these few days in Copenhagen we will be blessed or blamed for generations to come; we cannot permit the politics of narrow self-interest to prevent a policy for human survival.

Enter Barack Obama. It seems that only he can swing triumph from the jaws of disaster now.

The world needs a Churchill now, I write in the *FT*. For the people of Tuvalu today, and all the rest of us tomorrow, let it be Obama.

But he says nothing to advance the sham tabled by Hillary Clinton. The world needed him to seize his moment and show the political will of Churchill, taking the summit to a new place, shaming the pessimists and foot-draggers into silence, assuming greatness as a global leader forever. Instead he gives us a dose of Chamberlain. He wants to wave a piece of paper that will not get us on course for keeping the carbon enemy caged.

I am sure many would consider this an imperfect framework, Obama says. But we can either take a historic step forward, or we can choose delay and repeats of the stale arguments until climate change becomes irreversible.

The arguments from the United States, it should be recalled, have been staler than those of most governments in these negotiations, and for longer.

As for China, how its cautious leaders now let their people down. The Chinese economy is essentially resident on its coastal plain. Every percentage point of GDP the leadership proudly posts will end up destroyed by the march of irreversible climate change, and much more besides. The Chinese leaders know this. The Chinese Academy of Sciences tells them so. They could have shamed the Americans into meaningful action by committing to a cap on emissions within a few years, with steady reductions thereafter: the measure needed from them if we are to have a realistic crack at capping global warming at 2°C. They could have scared the climate-denying American heartland into low-carbon action by showing them, whatever their beliefs, that they will be buried economically by a tsunami of Chinese cleantech industries unless they act.

Beyond these two countries, and their 40% share of global emissions, there is a general shared responsibility of course. Picture this convention of world leaders as the board meeting of a giant corporation. The board has known they had to deliver a master plan for many years, with the very survival of the corporation at stake. And they turn up with no plan, bickering among themselves over trifling matters. Just imagine the shareholder reaction.

Finally, there are the hosts. The Danes and the UN have made a dog's dinner of this vital summit, from beginning to end, despite the long lead time they have had to prepare. They have treated negotiators, press and NGOs unforgivably. Ambassadors, correspondents, CEOs and campaigners alike have had to queue for hours in the freezing cold. Civil society has been shut out of the endgame with no meaningful representation. The paramilitary Danish police are already under investigation for excess in their treatment of protesters. The very cause of democracy has been set back at the Copenhagen summit.

At stake is a liveable future on the planet. Parents of enquiring teenagers the world over now face ghastly questions.

Dad, why did world leaders – acknowledging that our future is at stake, knowing that they needed to do something that could cap global warming below 2°C – leave Copenhagen with a piece of paper heading for 4°C? Why couldn't they even agree a binding agreement on that first step? Why did the rich countries find it so hard to help the poor countries, eventually conceding $200bn a year by 2020, when they can quickly stump up almost $10,000bn to bail out their banks?

Well darn, er . . .

As we digest the implications of our collective failure in Copenhagen, we surely have to think hard about capitalism in the form we have allowed it to evolve. The fact is that as things stand, there is no place on the global balance sheet for the assets most relevant to the survival of economies, ecosystems and civilisation. Meanwhile, there is plenty of space for spectres that we label as assets while shovelling their attendant mega-risks off the books.

How dumb is that? What an epitaph we are teeing up for ourselves!

I take the ferry back to the UK. The exhausted student protesters on the boat exude an air of collective dejection that is heartbreaking.

I do not think they will be in a mood to forgive anyone anytime soon.[10]

Chapter 10

As bad as the credit crunch

One by one the consequences of failure in Copenhagen began to play out. Within days of the summit, Eon and Centrica both said they would be less likely to build coal plants attempting carbon capture and storage. Where was the imperative for them to do so? Before the month was out, a coal producer, Suek, announced it planned to float on the London Stock Exchange. What did it have to worry about from climate regulators? In January, the first departures began from carbon-trading desks in financial institutions. Why stay in a shrinking market?

As one anonymous banker said to *Reuters* of the Suek flotation: 'There haven't been any good opportunities in this sector for a long time, and the sector is on its way up, so therefore this will be a positive story.'

Many more such 'positive stories' could be expected now.

The Copenhagen Accord had set a deadline for all countries to officially state their emission reduction targets or list the actions they planned to take to counter climate change. Before January was out even that small deal was dropped.

There was no doubt that 'Climate-gate' had been a big factor in the failure of governments in Copenhagen. It had dominated so much of the news coverage. Politicians genuinely seeking progress had been forced onto the defensive. Politicians looking for an excuse to escape from action found they had a workable one. Now, it was clear, the sceptics would scour every corner of the climate change world, looking to turn any misstep into 'proof' that global warming was a scam.

In February, they found one. The IPCC's 3,000 page fourth report contained a mistake: a misstatement about the retreat of Himalayan glaciers. All hell broke loose in the press again.

In the UK, polls showed the number of Britons who did not believe in global warming was rising. Most Conservative MPs, including six members of the shadow cabinet, were now sceptical.

***

Back in Washington, smarting from the catastrophe in Copenhagen, President Obama found new resolve on bank reform. His administration devised a draft regulation that they referred to as the Volcker Rule, after its chief architect, a former Federal Reserve chairman. Paul Volcker argued for separation of commercial and retail banking. He wanted to forbid any bank holding deposits guaranteed by the government from operating hedge funds or private equity funds or from trading on its own book.

Banking lobbyists geared up their massive resources to fight the proposed reform. In Davos, at the World Economic Forum, Barclays President Bob Diamond lashed out at Obama's nerve. 'If you say that large is bad and we move to narrow banks the impact on jobs and the global economy will be very negative', he warned.[1]

Diamond had other interesting observations to offer this annual convention of the business elite. He professed himself 'incredibly proud' that Barclays had survived the crisis without direct bailout money, and railed against the unfairness of a regulatory crackdown that punished successful, well-managed banks like his own as much as failed ones. 'I am angry at banks that had poor management and poor regulation', he said.

***

Through 2009, the Industry Taskforce companies had found themselves immersed in battling the recession that followed the financial crash, like everyone else in the business world. Yahoo, First Group and Foster and Partners pulled out of the group. Buro Happold joined. Arup took over the chairmanship from Virgin. By February 2010, we were ready to have another go at warning about oil depletion.

In the interim the oil price had sunk, amid the worst recession in a century, from its record high of nearly $150 a barrel to below $40. It had then rebounded to around $80 a barrel, high by historical standards, even though the world was still struggling with profound recession. Those of us worrying about early peak oil were pretty sure we knew what that was all about: underlying structural fear in the market about supply, whatever the bullish rhetoric from the industry.

The Taskforce released its second report at a press conference in the Royal Society. This time we wanted to be even clearer with our warning. The chairmen and CEOs from Virgin, SSE, Stagecoach, Arup and Solarcentury used deliberately stark language in the foreword. 'The next five years will see us face another crunch – the oil crunch. This time, we do have the chance to prepare. The challenge is to use that time well . . . Our message to government and businesses is clear. Act . . . Don't let the oil crunch catch us out in the way that the credit crunch did.'[2]

In this report we updated the evidence that defines global oil reserves and extraction rates, and concluded that the global peak production rate for oil would likely occur within the decade, very likely by 2015 at the latest – at a value no higher than 92 million barrels per day. This compared with the then record extraction rate of 87 million barrels a day set in July 2008. By April 2013 the average global production for a quarter stood at just over 90 million barrels a day.[3]

The fundamental issues identified in the 2008 report had remained unchanged, though the global recession had slowed demand for oil in the interim. The flow rate data showed that increases in extraction would be slowing down in 2011–13 and dropping thereafter. Given the long lead times involved in developing the necessary infrastructure, this trend would be unlikely to reverse within the next five years. The industry was not discovering new giant fields at anything like a sufficient rate. The concerns about the reserves quoted by the OPEC countries – critical as they are to the confidence levels associated with future production capacity – had not gone away.

Our summary of data showed global oil production dropping at 1% a year from 2015. If the then IEA forecast of demand rising to 105 million barrels a day in 2030 were to prove correct, supply would fall short in 2015. The Taskforce, however, feared that strong oil demand growth would persist in China and India. If that happened, then global supply would fall below demand before 2015.

Our concerns about China and India were rooted in predicted numbers of vehicles – even assuming fuel-efficiency improvements, and oil use by city dwellers, projected forward from existing use rates. Factoring in such calculations left us worried that demand levels could exceed 120 million barrels a day by 2020, just as the IEA had once itself forecast. That would give quite some shortfall, if we were correct about production not exceeding 92 million barrels a day.

Note that the UK Industry Taskforce analysis assumes no 'surprise' disruptions in the oil industry's confident plans for massive new production required of them in deep water and the Arctic, if they are to keep maintaining global supply. But the industry's record in recent years suggests this may be – how best to put this? – overgenerous.

We didn't make this point in our report. We should have. The record was already dubious. Two months after our report was released, the Deepwater Horizon rig, under BP's control in the Gulf of Mexico, exploded, killing eleven workers, creating the world's biggest ever oil spill.

Royal Society, London, 10 February 2010

At the launch of the second UK Taskforce report Richard Branson takes the podium to kick things off. This time the room is full of journalists. I sit on the stage with four other leaders of Taskforce companies, waiting my turn to speak. In the front row I see a senior civil servant, Chris Barton, head of International Energy Security at the Department of Energy and Climate Change. He has asked to respond on behalf of Her Majesty's Government, from the podium, as soon as we have finished. The Taskforce has willingly agreed to this.

I have been busy lobbying, ahead of this event. I have shown a draft of the report to Prime Minister Gordon Brown's right-hand man on energy, Michael Jacobs. Barton is probably here under sufferance, told to attend by Number Ten Downing Street.

Branson speaks for three minutes, cameras flashed constantly. Casually dressed as ever, he speaks in his trademark style, with a diffidence that belies his towering business acumen. It isn't until you meet the team he has assembled around him over the years that you get to appreciate the steely depth that lies behind him and his Virgin brand.

Oil is and remains the blood of our economic life, Branson says. The price of extracting oil is going up and up and up. We need to prepare for the peak.

He emphasises the Taskforce's key demand. The government must bring together business and the public sector to plan for the future and establish a clear contingency plan to combat a rise in the price of oil. We need to accelerate investment in renewable technologies, energy efficiency and alternative fuels. Nobody would be happier than those of us at the table today if our prediction did not come true, he concludes. Or better still had been mitigated by action taken from today.[4]

Branson sits down, and one by one I and the other company leaders offer our views.[5]

Now it is Chris Barton's turn.

Thank you for the compelling presentations, he begins. I suspect there may be many in this room who think that the government denies the possibility of an oil supply crunch, and that it does little to reduce the risk. This view is wrong. I have three points to make.

He speaks with a kind of polite assertiveness that hovers on the edge of belligerence. He reminds me more of a politician than a civil servant.[6]

First, there are two views of the future of oil supply and demand. On the one side we get people being called complacent and on the other we get them being called alarmist. So who is right? We don't know. But we do recognise that the risk of rising and volatile oil prices is real.

This, I think to myself, is an improvement on the Wicks Review then.

Second, we are taking action to mitigate these risks. We are doing many of the things that are recommended in the report today.

He elaborates. The government is making progress in energy efficiency, electric vehicles, renewable energy.

But. Third and final point. We need to work together to do more.

At this point, sitting on the stage facing the rows of journalists, I struggle to keep my face in order. Work together. That would have been good three years ago.

We may need to do more than we have planned, Barton continues. We are doing some of the things recommended in the report today, but we are not doing all of them.

Should we be? Well, we need to work together on that. Today's report suggests that the government set up a forum with industry to examine these issues. As it happens, we are in discussion with the Energy Institute about doing just that. We need to work out the details, but the aim will be to get a small group of experts from a range of different viewpoints to discuss practical solutions to the risks we face.

So we don't have a firm view on what the future holds for oil supply and demand, he concludes, but we do recognise the risks. We are taking action to mitigate those risks, and we want to work together to do more.

Three weeks later Energy Minister Lord Hunt and his officials met industry representatives at the Energy Institute in London for half a day. The existence of the supposedly private, behind-closed-doors occasion was leaked to the press the day before, and not by the industry side. Since the headline flagged that the minister would be 'calming rising fears' of peak oil, and a Department of Energy and Climate Change spokesperson gave a lengthy quote, we can safely assume both that the government leaked this news and that the minister would not particularly be in listening mode. The spokesperson played down the significance of the session. The government had always taken supply issues seriously, she said, and met different parts of industry on a regular basis. 'We do this all the time; it is just a normal stakeholder meeting.' There was no marked change in ministerial policy, she emphasised.[7]

I did not attend the meeting. I had a long-standing engagement to debate the nuclear industry. The minister barely attended it, turning up only for the last 45 minutes. My Taskforce colleagues reported that the engagement was a predictably frustrating exercise: a very long way from what we wanted. We sought a full national contingency plan, drawn up after extended study and consultation by the finest minds in the relevant ministries and companies. We wanted to see the government face up to the key questions about what to do should a permanent descent of global oil production shock the system. How would we keep supermarkets and shops stocked? How would we keep the nation moving? How would we go about the necessary emergency energy conservation? What would the rationing regime be? How could industry best help government? How would we accelerate the generation of our domestic non-oil energy? How would we maintain the coherence of civil society? How would we head off wars for the remaining oil supplies?

Three weeks after the meeting, it became clear that the long list of players not sharing the British government's sanguine view included the US military. A Joint Operating Environment Report of the US Joint Forces Command warned of massive oil shortages by 2015. The shortfall in global oil output by then could reach almost ten million barrels per day, they concluded. The report drew some stark conclusions: 'While it is difficult to predict precisely what economic, political, and strategic effects such a shortfall might produce, it surely would reduce the prospects for growth in both the developing and developed worlds. Such an economic slowdown would exacerbate other unresolved tensions, push fragile and failing states further down the path toward collapse, and perhaps have serious economic impact on both China and India.'

This warning of global depression triggered by oil supply problems, on the same time frame as the UK Taskforce feared, came with an obvious but nonetheless sobering observation from history: 'One should not forget that the Great Depression spawned a number of totalitarian regimes that sought economic prosperity for their nations by ruthless conquest.'[8]

***

At this point in the history, shale gas begins to feature as an increasingly important factor in the run of events. At the Energy Institute meeting, a Shell executive gave a presentation on the boom in production of unconventional gas in the USA, and offered the hope that this might become a major reason for reduced fear over oil depletion, in that transport could in principle be switched *en masse* to gas, and a glut of gas seemed to be emerging. Tony Hayward was pushing the same line at the time. In Davos he had described the US shale gas boom as a 'game-changer' for US energy supply.

For much of the history of the oil industry, gas and oil trapped in shale deposits had been considered unobtainable economically in commercial quantities. Shale is a low permeability rock quite unlike the permeable strata needed to host conventional gas or crude oil reservoirs. Shales had for most of the history of the oil industry been thought of only as 'source rocks' – bearers of the organic carbon needed to generate oil and gas – rather than reservoir rocks. By 1998, however, the industry's technologists, with the help of their Department of Energy subsidies, had found a way to access and produce in-situ gas and oil. By drilling horizontally and injecting fluids under pressure to force and hold fractures open, the hydrocarbons could be produced relatively cheaply and in volume. This process, known as fracking – short hand for hydraulic fracturing – soon began transforming the US energy scene. In 2000, only 1% of US gas production came from shale. By 2010, the figure was 20% and rising fast, driving the gas price down as it did so. Oil giants were piling into shale gas, buying up the smaller pioneering companies. There was also growing interest in producing oil the same way.

Only three years before, most energy executives had been worrying about the viability of gas imports from Russia and Qatar. Recall that in August 2007, Conoco's CEO had said 'the world has a gas problem'. This is how quickly narratives can turn around in the hype-laden world of energy.

The oil industry was seeing great opportunities to export the US shale gas phenomenon to other countries, and many countries – not least those dependent on Russian gas – were desperate for this to happen. The IEA suggested that shale gas might expand the quantity of economically extractable gas in the world by 50%. Licences for exploration were being taken out all over Europe by 2010, though no gas had been produced.

But with the apparently golden opportunities were emerging dark downsides. Fracking requires thousands of litres of water and chemicals, many of them toxic. The Bush administration had essentially deregulated the gas industry in the early years of the century in order to allow unfettered expansion of shale gas production. Fracking had been unleashed across America at scale, almost unmonitored. Now stories of contaminated groundwater, dire air-quality problems, and sick livestock and people were beginning to circulate. And the industry's assertion that gas production was less damaging in terms of greenhouse-gas emissions than coal was also coming under the microscope, with fears of large-scale methane leakage cancelling out the advantage.

The industry denied all this. But it knew there might at least be a problem. When Exxon acquired shale gas specialist XTO Energy in December 2009, the oil giant inserted a clause that would allow it to walk away if environmental regulation were to make fracking technology uneconomical.[9]

BP joined an industry rearguard action to keep the Environmental Protection Agency's hands off shale gas oversight. The company began lobbying on Capitol Hill for states to be given oversight of fracking, not the federal government's specialist environmental agency.

In the UK, the first exploration drilling for shale gas began in April 2010, to the east of Liverpool. Expectations were running high. The comforting narrative that shale gas was a game-changer was deeply appealing to many people.

Oxford University, 14 April 2010

SolarAid, the solar lighting charity set up with 5% of Solarcentury profits, is making promising progress in the four African countries it is operating in. It is time for me to take a trip into the world of philanthropy, and to talk to the kinds of funders we will need if we are to grow as fast as the SolarAid team dreams: foundations, venture philanthropists and all the others interested in how best to use the multiple billions that are given and invested at zero or low return in social good, even in the teeth of the worst recession in a hundred years. One of the very best places to do this is the annual Skoll World Forum for Social Entrepreneurship, in Oxford.

I drift around, meeting old friends and contacts and making new ones, pushing the boat out, making a mini-pitch here, having a catch-up there, doing the chairman thing.

On day two, I notice a session entitled 'The Neuroscience of Change: Understanding the Brain, Influencing Behaviour'. My long experience of trying to influence behaviour, when it comes to systemic risks and the societal response to them, is leaving me increasingly perplexed. I can't understand why so many arguments that seem so obvious to me, and many people like me, are so very dismissible, even ignorable, for so very many others. Maybe, I think, it might be a good idea to invest an hour and a half in listening to human brain experts talk about the human brain.[10]

This turns out to be good thinking. Or at least, illuminating thinking.

I go into this session believing, like so many science-steeped professionals of a certain age, that a good logical argument, compellingly assembled, can change minds and hearts, and even do so *en masse*. I come out suspecting that the reverse may be the case. That 'logical' is the last thing human thinking, individual and collective, is. That too compelling an argument can even drive people with a particularly well-insulated belief system deeper into denial.

Three eminent professors of neuroscience tell of an explosion of discovery in their subject in the last twenty years. The book they recommend tells the basic story in its title: *Predictably Irrational: The Hidden Forces That Shape Our Decisions*. Listening to them, I learn one new thing after another.

The great majority of thinking we do never even reaches our conscious brain. We have a nervous system deeply wired during human evolution that is in many ways unsuited to decision-making in the modern world. We are bad at projecting ourselves into different emotional states and imagining how different our decision-making will be in them, though it manifestly will. And so on.

Not everything I learn is a setback to my own psychology. I also hear about the discoveries of how important community is in the human mind. The unconscious brain is not a selfish thing: we tend to be pro-social in our thinking. Many humans yearn for community. The notion that there is no such thing as society, a belief of so many on the right of politics, seems to be far from the case. Certainly we are tribal in our thinking – much swayed by the 'in-group', as the professors call it, and less prone to empathy outside it. But selfishness and deceit are harder for our brains to do than empathy is. The experiments show that around two-thirds of us are pro-social. We like to share.

I leave Oxford feeling that some kind of veil has been lifted from my eyes. I have been to so many seminars in my life, but rarely have I encountered one that changes my perspective so.

On the train back to London, it occurs to me that I should no longer expect to win confrontational debates with the incumbency, whether its generals or its footsoldiers. I can be more chilled in those debates, feeling less that it is about 'winning' or 'losing', more about planting seeds in the minds of those listening.

And I realise that even though I was already interested in the role of community in fashioning a survivable energy future, I am even more so now.

In April 2010, the SEC charged Goldman Sachs with a $1 billion fraud at the height of the credit crunch. The lawsuit alleged they had wilfully mismarketed sub-prime securities to clients knowing them to be toxic, meanwhile working with a hedge fund to short the same securities. Recrimination was by now thick in the air on Wall Street, where it was widely suspected the floodgates would open on litigation. The IMF, professing that banks were more powerful now than they had been before the crisis, proposed tough new measures to cut them down to size, taxing both profits and pay.

Meanwhile, the IMF's own activities were not without implication. The debt-laden eurozone was looking increasingly shaky, with near-bankrupt Greece at the head of the sick list. A joint IMF and EU rescue package came with the strict condition of an austerity programme. The cuts imposed by the government as a result led to riots in Athens in which protesters petrol-bombed a bank and three people died in the blaze.

***

A steady drip of reasons for concern about peak oil followed the meeting between the British government and industrialists. ConocoPhillips CEO Jim Mulva confessed that in his view pursuing new oil reserves no longer paid: it was becoming just too difficult to access new sources. This seemed a strange thing for an oil boss to say. Coming from the USA's third biggest oil company, it certainly did not auger well for a concerted effort to make supply meet demand. Neither did everyone in the industry seem to agree with Tony Hayward's vision of shale gas as a game-changer. Chevron's CEO, John Watson, said his company would not be joining the rush, warning that the 'price tag is too high' to justify the investments.[11]

An *FT* columnist wondered if policymakers, economists and 'peak oilists' are 'starting to speak the same language'.[12] 'It's still the rare politician or industry executive who would use the phrase "peak oil"', wrote Kate Mackenzie, 'but in the UK, a country for whom domestic oil production decline is very much a concern, the issue has become almost mainstream.'

She wrote that the day after the Deepwater Horizon rig blew up, and the first of what would end up being 4.9 million barrels began gushing from the well, BP's safety analysis for this well had deemed such a spill 'virtually impossible'. By Day 22 of the disaster, oil was still pouring from the uncapped well and a horrified world was beginning to realise that BP had no immediate solution in sight for their virtual impossibility.

Chapter 11

You are the flip side of austerity

Sundance, Utah, May 2010

A hundred very senior executives gather in an out-of-season ski resort in the Rockies. They come from the venture capital community in Silicon Valley, the clean energy companies they have invested in, and – a big difference to a normal event of this kind – the energy incumbency companies whose markets they are endeavouring to displace. The subject of the two days of discussion is simple. Why isn't the cleantech revolution working?

I wake in a log cabin on the first morning, see forests and mountains leaping off to the west under a deep blue sky, and have to suppress an intense desire to play truant.

A leading light of the venture capitalist industry, Stephan Dolezalek of Vantage Point Capital Partners summarises the core problem.

Only eight VC firms have completed more than ten deals a year in cleantech in recent years, he observes. What have we learned? That this is not easy. That this is not like the internet. Capital costs are high and failures are inevitable, and plentiful.

I feel a small satisfaction that Solarcentury, a Vantage Point investment, is not one. Yet, at least.

Some portfolio companies are on the edge of a breakthrough, Dolezalek continues. But we badly need one huge success story to happen. So this is the Big Question. Should we keep going or should we ease up?

None of the senior incumbency players present argue in favour of Ease Up outright. But they offer eloquent and convoluted defences of the status quo: a catalogue of rationales as to why any major departure from how things are currently done is so very difficult in the energy industry.

BP is there, in the shape of BP Solar CEO Reyad Fizani. I give a presentation on Solarcentury's progress. The story is not bad if you like average annual 40%-ish growth for a decade, but disastrous if you are a VC, like Vantage Point, trying to replicate your successes with the digital and internet revolutions in cleantech.

I had no idea you had achieved so much, Fizani tells me in the break afterwards. He is not a VC.

He asks me what I would do if I were him, in his situation in BP. His frustration is palpable.

I ponder how to deal with this one. If I were him, I would never have joined BP in the first place. But I elect to answer as though I were in his shoes, and in his mind not my own.

Go to Bob Dudley, I say, or whoever replaces Tony Hayward when he gets fired after the mess is cleaned up in the Gulf, ask for the budget to buy every decent downstream solar company in multiple countries, and create a global solar colossus. The new CEO will be under massive pressure to repair the BP brand, and a bit of serious solarness might be just the right kind of delivery vehicle. And of course it would be a smart business idea, given solar PV's proximity to price parity with conventional electricity in multiple countries.

I would hate that outcome, I reflect, were he to action it and Solarcentury forced to be part of it. VCs own the majority of shares in the company I founded. I wouldn't have a say.

But I'm pretty sure that I have nothing to fear when it comes to BP ever being serious about this idea. They have a recarbonisation strategy, not a decarbonisation one.

And I can see from Reyad Fizani's reaction that he thinks the same.

BBC TV Centre, London, June 2010

Day 51 of the Macondo spill and the oil is still pouring out. Every measure tried to date has failed to stem the flow. BP releases its annual Statistical Review of World Energy as scheduled. I have persuaded the *FT* to run an op-ed the same day. As BP now admits, it did not have the tools to contain a deep-water oil leak, I write. Its failure with that risk must now raise questions about its approach to other risks. Top of the list must be the threat that global oil production will fall sooner than generally forecast, ambushing oil-dependent economies with a rapidly opening gap between supply and demand.

Newsnight, BBC TV's flagship evening current affairs programme, are interested in this argument. I am called into the studio to be grilled by famous attack-dog anchorman Jeremy Paxman. The interview is live, lead item on the show. BP has turned down the invitation to field someone. Instead, a Norwegian oil company CEO is there to explain what rubbish I am talking.

Paxman has verbally eviscerated a long catalogue of people on live television. I have rarely been as nervous.

I console myself by thinking that no matter how badly I might do, I am unlikely to be as disastrous as Tony Hayward has been in some of his interviews in the Gulf of Mexico. On Day 25 of the disaster he observed that the spill was relatively tiny compared to the very big ocean.[1] On Day 42 he complained: 'There's no one who wants this over more than I do. I want my life back.'[2] As I watched that one on TV, thinking of the eleven people who had lost their lives – people whom Hayward had momentarily forgotten in his self-absorbtion – I actually felt sorry for him. This, I knew instantly, was an elephant of a gaffe, a mistake that would echo down the history of corporate PR. All PR.

As Paxman asks me his opening question, I wonder if my heart rate is survivable, whether anyone has ever dropped dead on air in this studio.

I have been media-trained, though, and I know the importance of controlling your breathing. I get out as punchy an answer as I can.

Of course you would say that wouldn't you, Paxman says as he turns away from me to get a reaction from the oil boss, you sell solar panels.

> By the time he finishes this effective accusation of dishonesty, he has his back to me.
>
> I have a millisecond decision to make: let it go, or not.
>
> You'd have to think I am crazy to be motivated that way, I say to Paxman's back. I set up my solar company *because* of my concerns about fossil-fuel dependency.

The day after the Paxman interview, *Telegraph* writer James Delingpole raged against the gullibility of *Newsnight* for running a 'scare story' on peak oil. In *Newsweek* the same day, author Tom Bower called peakists 'charlatans'.

But this kind of ever-present frothing did not stop the worrying news. Lloyd's of London produced a report warning of catastrophic consequences if companies failed to prepare for peak oil. Former UK Chief Scientist David King sided with the scaremongering charlatans. When he had been in government, with Tony Blair's ear, he had dealt with the peak-oil issue by telling the Association for the Study of Peak Oil researchers to go away and come back when their findings had been published in peer-reviewed journals.

BP was finally able to stop the flow of oil from the Macondo well on Day 68 of the disaster, 16 July 2010. Within five days the company had published the results of an internal enquiry that absolved itself of sole blame for the spill. Three days after that, the board fired Tony Hayward.

It would be far from the end of any involvement in the spill for him. In Washington, the FBI and other federal agencies assembled a 'BP squad' for a criminal investigation. Shell declined to rule out legal action against BP for causing the post-spill regulatory crackdown that was slowing the whole oil industry offshore.

The banks were undoubtedly loving the extent to which the focus on Big Oil was now diverting attention from them. In July, Goldman Sachs settled with the SEC for a $550 million fine. It cost them less than a week's worth of trading revenues to escape serious criminal charges.[3] Cynics wondered how many days of oil revenues it would end up costing BP to escape the same.

***

In June, President Obama used an Oval Office TV address to call for a national mission on clean energy. Here he doubtless felt he was working with the grain. New renewable power capacity had topped fossil-fuel and nuclear capacity again in both the US and Europe during 2009. In Germany, the rapid progress with renewables led the German Environment Agency to conclude that national electricity could be 100% supplied by renewables by 2050, with wind and solar doing the lion's share of the work.

Modelling studies had long since suggested that modern economies like Germany's could be run by renewables, and much quicker than 40 years hence. One had shown that the entire global energy demand could be run by renewables by 2030. I shall return to this optimistic view in Part II.

Solar energy continued to grow, despite all the setbacks. President Obama proudly announced that he would be putting $2 billion of stimulus money into just two US solar companies.

In Britain, solar energy had yet to find a place on any government priority list. But it could no longer be completely ignored.

New Delhi, July 2010

The bus weaves miles across the far fringes of Heathrow's tarmac, with CEOs and chairmen clinging to straps like tourists being driven to a charter flight. We reach a souped-up portacabin labelled the Queen's Suite and, after the quickest check-in I have ever experienced, board a BA jumbo jet, chartered by Her Majesty's Government for their trade mission to India.

We land *en route* in Istanbul, and there pick up David Cameron. As we wait for him, I chat to Vince Cable, the Business Secretary. I bring up peak oil. He tells me he doesn't believe the below-ground arguments on supply constraint, and is much more worried about above-ground factors. I remind myself that he was once Chief Economist at Shell.

Flying over Iran, I stand talking to the former Chief Scientist at the Ministry of Defence. He assures us Iranian anti-aircraft missiles can't reach this high. If they could, and they fired one, they would make one hell of a story. A prime minister and half a dozen ministers are aboard, not to mention 29 captains of industry. Or perhaps more accurately, 28 captains of industry and me.

Delhi. A meeting at the Indian Chamber of Commerce. Almost the entire UK business delegation sits across a very long table from just four Indian civil servants. We hear the most senior Indian describe his government's plan for 10% GDP growth every year until 2020. It is painstaking, impressive in a way.

If you do that, I sit thinking, and China does something similar, and you keep going after coal the way you are, then you compound the greenhouse gases the rich nations have already pumped into the atmosphere and we lose a liveable future. It's as simple as that.

But the discussion focuses on how the two nations might collaborate, to mutual benefit, in the search for the 10% growth. This mission is all about access for UK plc to the mass markets of the rising Indian middle class.

There are several bank bosses in the room. The Barclays CEO, John Varley, launches into a pitch for the Indians to open up their financial services markets to British banks.

I study an ink mark on the table in front of me. I wonder what would happen to me if I said out loud what is in my mind.

Do you think they are *crazy?*

I risk a look at the Indian civil servants. They are doing fairly well as diplomats, but nonetheless unspoken words are written all over their faces.

Do you think we are *crazy?*

There is something post-colonial about this whole exercise. I can feel the resentment just below the surface in almost all the Indian officials and business people I talk to.

In a small meeting on solar, it surfaces. The Indian environment minister, Jairam Ramesh, and his British equivalent, Greg Barker, lead a condensed discussion. Half a dozen or so business people on each side are able to make short statements. I am one of them. The ministers then sum up.

Greg Barker does the best job he can, in the circumstances.

Jairam Ramesh has a reputation for being blunt. He ends the meeting by telling Barker that he shouldn't be under any illusions that the UK's colonial past gives us any special rights in access to Indian markets.

We get delegations coming here all the time, Ramesh says. He names a few countries that have been in town recently.

> You have to tell us what your unique selling points are, he continues, with a didactic edge. I haven't heard any. Other than perhaps this gentleman's solar roof tiles.
>
> He points at me.
>
> I could quite see those on the rooftops of Indian cities, he adds.
>
> I counsel myself against getting excited. I have seen enough to know what Ramesh is probably thinking.
>
> So long as they are made in India.

The third anniversary of the credit crunch in August 2010 and the second anniversary of the financial crash in September saw an explosion of reviews. It was clear that many in the commentariat considered the minimal financial reforms achieved in the interim utterly inadequate. 'Why it is still bank business as usual', an *FT* editorial declared.

Barclays marked the occasion by appointing Bob Diamond, 'the quintessential "casino" banker' as the *FT* saw him, as their new CEO.[4] Diamond hit out immediately at those who spoke of 'casino banks'. Such critics didn't understand how banks work, he said.[5]

They didn't? The subtext of the *FT* editorial read: 'Change depends on the ability and courage of regulators.' The regulators had failed, the *FT* argued.

The CEO of the Financial Services Authority at the time of the credit crunch and the financial crisis was Hector Sants. In 2012, he left the FSA and joined Barclays.[6]

***

The American military had offered its view on oil depletion. Now the German army added theirs. *Der Spiegel* ran a story on a study, not meant for the public, by a think-tank in the Bundeswehr. 'A permanent supply crisis threatens – and just the fear of it can cause turbulence in commodity markets and stock exchanges', the military authors concluded. 'The topic is so politically explosive that it is remarkable when an institution like the army uses the term Peak Oil in public.' The report majored on the probability of bottlenecks in the supply of important goods, including food supply, and spoke of partial or even complete market collapse. 'An alternative would be conceivable: government rationing and the allocation of important goods or the setting of production schedules and other coercive short-term market-based mechanisms in times of crisis.'[7] Democracy could even come under threat, the authors concluded.[8]

Here writ large was why the UK Industry Taskforce on Peak Oil and Energy Security wanted to see the British government act on the risk, and at minimum draw up a contingency plan with industry.

And now came evidence that the UK government was far more worried about the problem than it had ever come close to admitting, and that military officials were involved in their thinking. Peak-oil researcher Lionel Badel used the Freedom of Information Act to uncover the fact that DECC staff had held a meeting in 2009 with the Ministry of Defence and the Bank of England among others to discuss the risk. A ministry note of that meeting warned: '[Government] public lines on peak oil are "not quite right". They need to take account of climate change and put more emphasis on reducing demand and also the fact that peak oil may increase volatility in the market.'[9]

The meeting had taken place twelve months previously. Nobody had told the Industry Taskforce. A letter in response to Badel's Freedom of Information request, written by DECC officials and dated 31 July 2010, said the government could only release some information on what is currently under policy discussion because the discussions are 'ongoing' and of a 'high profile' nature.

In the wake of the Wicks Review, this was all news to the Industry Taskforce.

The note of the meeting a year ago suggested officials stick to a party line that the 'International Energy Agency is an authoritative source in this field'. It emphasised how the IEA believes there are sufficient reserves to meet demand until 2030, as long as investment in new reserves is maintained.

Slough, UK, September 2010

An opening ceremony with a difference. Solar roof tiles, glistening on the roofs of eight homes as though rained on under a grey autumn sky, can provide all the electricity the homes use, with plenty left over to power the battery car that the families living in the community share. Four technologies are each capable of heating the whole of the airtight, triple-glazed development: air-sourced heat pumps, ground sourced heat pumps, a biomass boiler, and solar thermal panels.

SSE, the British utility and investor in Solarcentury, has built this development for some of its workforce, as a showcase for what could be done. Half of Britain's carbon emissions come from buildings. Half of those come from residential buildings. These homes use no coal, no gas, no oil, no nuclear. They go beyond zero carbon.

Ian Marchant, SSE CEO, gives a speech.

This is a stunning glimpse into the future, he says. These homes took just eight months from the start of building to occupancy.

Chris Huhne, Secretary of State for Energy and Climate Change, follows him.

Low-carbon energy is going to be a massive industry, he says. Congratulations to all.

Back in London, I have a meeting at GE with Mark Elborne, their UK CEO. We have another demonstration project to talk about: solar schools, a flagship within GE's sustainable cities programme. We are destined to hit the front page of GE's website: a £50 million minnow partnered with a £100 billion giant.

I recount my morning with SSE to Elborne with a pride I can't disguise.

Just wait until we marry this kind of technology up in a few years' time with the storage technologies we are working on, he says. Then we'll see some changes in energy markets.

This is a good day at work.

Back home in Holland Park, I am dragged back to earth. A meeting of my flat-dweller neighbours convenes. We have all been thrown in purgatory in the past months as the owner of the mansion next door builds an underground cinema in his back garden. Platoons of Polish builders troop in and out of the mansion. Our flats shake all day in the incessant jackhammer hell they create. The owner of the mansion has moved out for the duration. We, his neighbours, can't.

He is an investment banker. He is spending his bonus, or part of it.

It took until September 2010 for global regulators to set banks a new capital rule. The Basel III agreement required them to hold 7% of capital in reserve against losses, up from the previous 2%.

It wasn't enough to stem criticism. A 7% capital ratio still assumed a bailout if a bank ran into difficulties, critics pointed out. 'We must press on with the breaking up of the banks', wrote the *FT*'s John Kay. 'Pledges of co-ordinated global reform have proved empty.'

The complexity of the web that the banks had woven in the run up to the 2008 crash was becoming ever clearer. Two years after the collapse of Lehman, fully 800 staff were working to unravel its affairs. And in October came evidence of systemic paperwork errors on some of the $1.3 trillion in mortgage-backed securities, so opening up a massive door for loss-making investors to launch their lawyers against the banks. It was clear that the legal profession would have guaranteed work very far into the future.

As for the rolling implications of the financial crash itself, the IMF's World Economic Outlook admitted that the world economy was stuck in near depression. It painted a picture of southern Europe slowly suffocating in debt, and left little doubt that fiscal tightening would trap northern Europe, Britain and America in their respective slumps for a long time.

Growing numbers of commentators were questioning the use of austerity in fighting the downturn at this time. As Nobel prizewinning American economist Joseph Stiglitz put it, 'to choose austerity is to bet it all on the confidence fairy'. What you had to do in a near-depression, they said, was borrow more to build infrastructure to create jobs: a modern version of the New Deal of the 1930s. And since this was the twenty-first century, let it be a Green New Deal.

In the UK, a group of green economics advocates set up by Colin Hines had been busy promoting a Green New Deal even before the 2008 crash. By 2010, it was getting some traction, with local governments looking to raise bonds for energy-efficiency projects to get people back to work.[10] But central government remained determined to maintain course with austerity.

***

In October 2010, Coal India, a huge government-owned coal company, offered 10% of its shares to investors at home and abroad in the largest ever initial public offering on the Indian stock exchange. What was at stake was essentially a $35 billion bankrolling of enhanced global warming by the capital markets. Yet Coal India's prospectus, crafted with the help of a clutch of big-name investment banks, did not mention climate change once in 510 pages of exhortation to invest. And invest the fund managers did, unfettered by risk regulation or any meaningful requirement to place a value on the climate consequences of their scramble for short-term profit. The offering was oversubscribed fifteen-fold, and the stock soared on the first day of trading, 4 November, valuing Coal India at $49 billion.

Those who ended up owning stock included some 484 foreign funds, 195 mutual funds, 44 insurance companies and many banks. Many of these investors were using ordinary citizens' money, and this would have included the nest eggs of many people worried about global warming and its dire impact on the world by the time they retire. But those people are mostly allowed no say in where their pension funds, insurance premiums and banking deposits are invested.

Enel, the Italian energy giant, had floated its renewables arm on the same day as Coal India. But Enel Green Power's initial public offering (IPO) was a flop. Institutional shareholders were particularly unsupportive. Some cited long-term fears for the viability of renewables generally. Again, this spectacle involved fund managers investing and trading for people who, given a say, might well have chosen to factor climate change into the decisions taken.

I wrote about all this in the *Guardian*,[11] and sent an e-mail alert to more than 2,000 key influencers. I received an electronic postbag in response that was an eye opener. There were a lot of capitalists out there, it seemed, who by now had grave misgivings about the course of modern capitalism. One message, from a venture capitalist, read as follows: 'This really echoes. "How much more suicidally dysfunctional can modern capitalism get?" It's out of control. And as we all know, it's the innocent people usually who end up paying the price.'

***

BP's bill for the Macondo disaster now stood at $40 billion and counting. In Bob Dudley's very first speech as CEO he hit out at 'fearmongering' over the Gulf oil spill. But genuinely scared scaremongers were active all over the capital markets and the energy markets as the first year of the second decade of the twenty-first century neared its close.

Some of them, as I knew, were officials within the International Energy Agency. And as the UK government had said, much would hinge on what this agency had to say for itself when it came to peak oil.

Could the concerned officials in the agency persuade the officials able to be unconcerned, or willing to play politics, to issue a strong enough collective warning that anyone would take any notice?

International Energy Agency, Paris, 19 November 2010

I have been hoping to meet Chief Economist Fatih Birol, he of the enigmatic statements about peak oil. Instead I get one of his juniors, Trevor Morgan. At least we meet in Birol's office. I can say I have seen his chair.

I approach Morgan using the language of risk, as I have taken to doing since my mind-altering exposure to the neuroscientists in Oxford. I ask how much weighting the IEA gives the uncertainties in their *World Energy Outlook 2010* New Policies scenario, which is essentially their forecast for what is likely to happen: oil supply rising until 2035.

Morgan tells me they recognise downside risk but think there is actually more upside potential.

I say that the Industry Taskforce I am representing has fears for all the sectors of their scenario. Existing fields may deplete faster than they think. Reserves may be overstated by national oil companies. Those fields coming into production might meet delays, for example deep-water regulation, security in Iraq and so on. Oil yet to be found might be less than they forecast. The gas story, and hence natural gas liquids and gas-to-liquids, might be less compelling than expected. Unconventional oil in the tar sands and elsewhere may be slowed by policy backlash or sheer operational difficulty.

He argues the toss on every point. Supergiant oilfields like Ghawar in Saudi are hardly likely to behave like Cantarell in Mexico: the IEA has an employee working for them who knows Ghawar well, so they have few fears of a collapse there. And so on.

I ask whether he ever worries about the difference between what they were saying in 2005 and what they say today. They said in 2005 that crude oil supply would grow for a long time into the future. Today they say crude oil hit a peak in 2006 that will never be exceeded, and all the growth comes from unconventional oil. Didn't that make him just a tad insecure?

No, he says, and the IEA feels they are often misrepresented: what they said in 2005 wasn't that different from what they are saying today.

I ask him whether, if he were government, he would argue for a contingency plan for an oil crunch.

No, he says. The risk of early peak oil is too low to be worth the effort.

Downing Street, November 2010

I and seventeen other founders of small and medium-sized companies have been invited at quite short notice to Number Ten to have breakfast with David Cameron and George Osborne.

The Prime Minister explains the purpose of the next hour and a half. You can generally see when politicians are going through motions deemed necessary on the day. This breakfast is not one of those occasions, I quickly see.

My government has regrettably had to embark on a course of austerity to bring the UK economy back under control, he says. Our idea in bringing you here today is to understand what we can do to help companies like yours grow, and hence create more jobs, so helping to countervail the job losses created by our effort to get public spending under control. In our view, you are the flip side of austerity.

I wonder if anyone else is thinking what I am. That surely the government, ahead of deciding on the austerity course, would have had many ideas of its own about the flip side.

The Prime Minister asks the attendees quickly to introduce themselves, and their companies.

They do so, and my heart sinks a little. One man outsources nurses for the National Health Service. A lady designs handbags. Another man assembles vintage cars. I quickly realise this is not Silicon Valley UK, it is much more Alan Sugar UK. Surely this is not the best cross-section of UK growth companies that Number Ten officials could find for their Prime Minister and Chancellor?

We move to the issue of where companies need help, and my heart sinks further. A couple of the people launch an appeal for less stringent maternity laws. Another raises capital gains tax, which he thinks should be cut. A woman describes at some length how she has worked hard to grow her company from nothing, and really feels that she deserves a tax-based reward for doing so.

I study Cameron and Osborne. Their body language tells me in tiny ways that they do not think this is going well. The people in the room are supposed to be the kind of people who must succeed if their model for the UK economy is going to have a chance in the next few years.

I get my chance to contribute.

I opt for the no-holds barred tactical boasting others have been using. I am founder of the UK's fastest-growing energy company this century so far, I say, active in one of the fastest-growing markets in the world, counter-intuitively – given our cloudy British skies – in solar. We are a British manufacturer. We are on the front lines of the green industrial revolution that the PM rightly sees as such fertile ground for job creation as the UK rebuilds. From where I sit, one of the key areas needing reform is finance. As I discussed with the Chancellor earlier, we must find ways to get the banks to provide credit for small and medium-sized enterprises. The financing of cleantech is in trouble all across the capital spectrum. Venture capitalists are in difficulties, and equity investors generally see no carbon risk in investing in fossil fuels and no carbon incentive to invest in cleantech. We must find ways both to fix this and to mobilise the peoples' money in cleantech investment, in the way Triodos Bank has started to do. The solar feed-in tariffs are providing a great opportunity to do this. The way that communities are creating their own solar projects, and financial institutions like GE and Triodos are coming to the financing party, is a real standard bearer for what you call the Big Society, Prime Minister. The opportunity to create energy service companies and green bonds should be a core activity of the Green Investment Bank that you are setting up.

Others pile in on the finance question. Cameron asks which banks were holding things up. One person who has supported me then prevaricates, unwilling to name his bank. So I recount how and why Solarcentury has fired

HSBC: because they aren't willing to invest in feed-in-tariff-backed projects that are safer than a government bond. We need to engineer the return of Captain Mainwaring in banking, I say, not stick with the current casino madness.

In his summary, the PM says his take-away messages included the need to clean up finance, and to create a culture of entrepreneurship.

My take away is that if these 18 SMEs and their founders are the best that UK plc can deploy, then the Cameron–Osborne project is heading for the rocks.

As we file out of Number Ten, it strikes me that watching Cameron and Osborne in action has been like reviewing a certain kind of undergraduate project. I saw many of these in my days as an academic. Bright and enthusiastic students, usually from public schools, who are good at talking, but who haven't really done much preparation work at all.

Chapter 12

Houston, it's just possible we may have a problem

Entering 2011, the oil price was pushing its way up towards $100 again. The IEA duly warned that the rising price would threaten the fragile recovery under way in the industrialised nations. Rising oil prices tend to drive up coal and gas prices as well. The US shale gas boom was proving to be an exception, in that the US gas market is isolated enough from other world markets that rapidly expanding American shale gas production could drive domestic gas prices down, while at the same time prices in other markets rose. In the UK, they were rising high enough to be a political issue. The Major User Energy Council warned that rising energy costs were damaging the British economy. As for consumers, the average annual combined gas and electricity bill reached £1,230. The number of households in fuel poverty was soaring.

At this point a major new reason for concern about oil supply crystallised in Saudi Arabia, one that had been brewing for a while. A senior official warned that if current domestic oil-consumption patterns persisted, the kingdom would be burning most of its oil production in-country in less than twenty years. 'Oil exports and economic growth will be constrained if there is no mix of alternative energy', said Hashim Yamani, a former commerce and trade minister. 'We won't be able to leverage prices of oil to build our institutions.' On current trends – mostly driven by rapidly rising use of oil in electric power plants to meet soaring electricity demand – the world's largest oil exporter would need eight million barrels a day by 2028, roughly equivalent to its current production, merely to meet domestic energy needs. Solar and nuclear energy would have to be fast-tracked to slow this trend, Yamani said. 'We will not just buy or import institutions and sell energy. We have to develop industries and research related to alternative energy.'[1]

Yamani's alarming perspective fascinated me on two fronts. First, like Sadad al-Husseini, he seemed to have little confidence that the Saudis had so much oil that all they needed to do was go and open up some new fields. Second, the US shale gas boom – touted as a panacea quite widely by now, if it could be replicated in other countries – did not feature in his world view.

Neither did it seem to be featuring in BP's, at least operationally. Bob Dudley's first big play as new CEO was to attempt a joint venture with giant Russian oil company Rosneft, hoping in so doing to leverage better access to Arctic oil. Had the deal gone ahead, it would have left Rosneft – and hence the Kremlin, essentially – as BP's single biggest shareholder. It didn't go ahead, because Dudley's old sparring partners in the joint venture TNK-BP filed for an injunction to stop the new venture, arguing that their shareholder agreement committed BP to the TNK-BP partnership as BP's primary vehicle in Russia. BP accused the oligarchs of holding the company to ransom. It made no difference. A judge in London put the Rosneft deal on hold.

The UK Industry Taskforce on Peak Oil and Energy Security continued to find it impossible to engage government. Energy Minister Charles Hendry had told me in November 2010, at a meeting in Parliament, that his chief scientist was conducting a review of views on peak-oil risk from a range of British experts. But whatever the outcome, he made it clear that the coalition government, like the last Labour government, would not work with us on a contingency plan.

In January 2011, Hendry sneaked out the results of the chief scientist's survey in a written answer to a parliamentary question. It read: 'In 2010, DECC's chief scientist sent out a call for evidence on the prospects for future oil supply to a range of experts. A number of responses received argue that a supply "crunch" (a tightness in the oil market), if not a peak in oil production, is very likely before 2020.'[2]

With this came no announcement of contingency planning. I wondered at the mindset of the officials who could combine this evidence of opinion among British experts with the worries coming out of Saudi Arabia about domestic consumption, and decide that they did not have a duty to act.

On the last day of January, the oil price passed $100.

Imperial College, London, March 2011

A day back in the university where I taught for eleven years. I have been persuaded to give a lecture to a vast class of first year engineers on the state of the world as I see it. Climate change, resource depletion, financial markets, the works.

Not much has changed in the class dynamic, in all these years. A few seem interested, engaged. Most seem elsewhere, in their heads. Many study mobile phones as I speak.

I run through my stump Powerpoint. I allocate half the time to the depressing stuff, half the time to what I portray as the building blocks for renaissance. It is vital to do that.

Time for questions.

A student who has asked several probing questions comes back with another.

When we're running the world in twenty years, he says, we'll find your generation will have messed it up completely and not left us with enough money to deal with the mess. That's the awful truth, isn't it? Why don't you just admit that? And apologise.

Many eyes are raised from mobile phones at this.

I ponder. I admire his spirit. There is a professor in the room checking that I'm not telling the kids anything too outlandish.

No, I say. I think it's probably worse. First, your generation will be running the world in less than twenty years.

I explain about the average age in the oil industry, and the average retirement age. 49 and 55.

And we'll all find, if people like me are correct or even halfway correct, that the world will be very messed up, well before twenty years are out. You'll be in the rebuilding business. And you will need every skill they can teach you here. And more besides.

And sure, I'm happy to apologise. I do so unreservedly. On behalf of my generation. For what it's worth.

The questioner is smirking. Nobody is looking at mobiles.

> Some of us have tried to stop what has transpired, is transpiring. I myself have been doing so for the entire two decades of your young lives. But we have failed, so far. I wish it weren't so. But it is. I'm genuinely sorry for that. They troop out to whatever is next for them. How to build bridges to nowhere. How to route pipelines that can never be filled. How to build nuclear power plants that don't suffer meltdowns. Whatever.

On 11 March 2011, the Fukushima Daiichi nuclear power plant was disabled by a giant earthquake and the deadly tsunami that followed it. The disaster that unfolded in the following months provided multiple windows into risk and the way individuals, institutions and nations handle it – before a crisis, during it, and after it. Let me take some snapshots.

Tepco, the utility running the plant, had declared that there was zero risk of multiple meltdowns. The enculturated complacency that sprang from this allowed plant managers, for example, to fill spent fuel ponds well beyond their design capacity, and to omit mandatory inspections, so massively compounding the radiation released when the cooling water evaporated in the ponds.

The Tepco managers at the plant held back from pumping seawater onto the overheating reactors and spent fuel ponds, losing vital time, because they feared destroying the value of company assets.

Japanese Prime Minister Naoto Kan raged at Tepco for withholding information from his government, it later emerged. There was a phase in the disaster when he thought he would have to evacuate Tokyo, and even that Japan itself would be lost. Post-mortem investigations showed that throughout the disaster there had been an enculturated tendency to cover up its enormity.

The disaster also showed extremes of human reaction at the individual level. On the one hand, after he saw his belief system going up in radioactive smoke, the president of Tepco retreated to his office and didn't emerge for a whole week, attending no meetings of the crisis management team, all without formally transferring responsibility to a deputy. On the other hand, some heroic workers returned to the plant to battle the disaster even when given the chance to be rotated out of the danger zone.

Elsewhere in the world, the reactions of nuclear advocates to the disaster as it unfolded seem stunning, in the rear-view mirror. On Day Seven, EDF

UK CEO Vincent de Rivaz said 'my determination to press ahead is undimmed. What Britain needs is nuclear.'[3] The exhausted workers at the plant had yet to get pumps working to begin the process of flooding the reactors. De Rivaz could have had no idea at all whether reactor meltdowns could lie ahead or not.

In the UK, desperation to keep the idea of a 'nuclear renaissance' on course led officials at the Department of Energy and Climate Change to plan a cover-up as early as Day Three. They held a meeting at which they planned how to bring the big nuclear companies in and plot how to spin the story to the British public. They, like de Rivaz, could have had no idea at all what lay ahead on the Fukushima site.

Pro-nuclear green commentators were amazingly quick to come to the defence of their belief system. British columnist George Monbiot declared as early as Day 11 that he had developed a love for nuclear based on the disaster thus far, because of how little damage it had caused. He could have had no idea what the worst case outcome of the disaster might be at that time.

The huff and puff of the post-mortems was already well under way in the commentariat months before the eventual 'cold shutdown' at the stricken plant. Radiation isn't that bad for you. Oh yes it is. It takes a once-in-a-thousand-years tsunami to cause a nuclear disaster. Oh no it doesn't.

In terms of defining the future energy of nations, however, perhaps the most important development took place on Day 16, far away from Fukushima. In the wealthiest German state of Baden-Württemberg, the anti-nuclear movement kicked the governing pro-nuclear party out of office after 58 years of rule. German Chancellor Angela Merkel had a potentially existential political problem on her hands. She rightly blamed her party's defeat on the Fukushima disaster, and she did what almost all politicians would. She changed her mind.

After a meeting with the leaders of the country's states on 15 April 2011, she turned her back on her pro-nuclear past. 'We all want to exit nuclear energy as soon as possible', she said, 'and make the switch to supplying via renewable energy.'[4]

A significant proportion of German industry immediately came out with her. The Association of German Energy and Water Industries (BDEW), representing some 1,800 utilities, had until this point been one of the more reliable pro-nuclear voices. In early May 2011, however, it called for the 'swift and complete' abolishment of nuclear power, by 2020 ideally but 2023 at the latest. Only the two biggest nuclear operators, Eon and RWE, opposed the decision, and it wouldn't be long before they halted their nuclear plans in both Germany and the UK.

Germany was *en route* to a future without nuclear power stations. Merkel's decision was ratified at the end of May.

On the progress of the German quest for energy transition – the *Energiewende*, as it was called – much would come to hinge.

***

In early 2011, the WikiLeaks revelations were sweeping the global media, day after day. There seemed no end to the stunning insights into the ways that governments operate out of the public eye. In February, oil came under the spotlight.

One cable showed that a US diplomat, reporting back to Washington from Riyadh, had been convinced by ex-Saudi oilman Sadad al-Husseini that the kingdom's potential reserves had been overstated by nearly 40%, and that the world's number one producer was 'running to stand still' as a result of operational challenges. 'al-Husseini is no doomsday theorist', the cable concluded. 'His pedigree, experience and outlook demand that his predictions be thoughtfully considered.'[5]

Another passage recommended that the US help Saudi Arabia turn itself into 'the Saudi Arabia of solar' as a way to overcome their growing oil supply problems, and keep the lights on for their fractious middle class.[6] In relaying this message, American diplomats came across like Silicon Valley visionaries on the role renewable energy industries could have in powering the world a couple of short decades hence.

The reasons for their concern would have been theoretical at the time they wrote the cables. But by February 2011, these concerns were playing out in reality across the Middle East. The 'Arab Spring' was under way. Anti-government protests that began in December 2010 in Tunisia were spreading into Egypt and beyond. Saudi Arabia watched nervously as these protests spread to neighbouring Bahrain.

In Libya, as rebels strove to overthrow President Gaddafi, a small but significant portion of global oil supply was shut off. Saudi spare capacity couldn't replace it, like for like. The Libyan oil was light sweet crude. The Saudi spare capacity was sour oil, requiring quite different refining. Suddenly the world faced another oil crisis. Oil could hit $200 a barrel, investment bank Nomura warned.

Saudi Arabia needed to lift its oil production to allay price jitters and to calm its population, the better both to control the global oil price and to stay in power. Facing a 'day of rage' protest in Saudi Arabia in March, King Abdullah threw $36 billion to his people in handouts of different kinds.

It failed to woo the activists. The Saudi government duly mobilised 10,000 troops in their north-eastern provinces. This was looking very bad. Fears that the Arab Spring contagion would spread to Saudi was now pushing the oil price up and stock markets down. 'The world's economic fate now hangs on the success of Wahabi oppression', wrote Ambrose Evans-Pritchard in the *Daily Telegraph*, with accurate candour.

On 12 March, the Saudi day of rage fizzled out under a blanket of high security. But the very next day the Bahrain government asked Riyadh for help after their police force was routed in a pitched battle with protesters. The day after that, the world witnessed Saudi tanks rolling into a neighbouring state. For many observers, it looked as though Western governments had come to a private agreement: we take out Gaddafi, and you can do what you want in Bahrain.

Two days later the Bahraini protesters were disbanded with deadly force.

On 5 April the oil price hit $120, its highest ever value in sterling, given the devaluation of the pound since the $147 record oil price in June 2008. The world economy was once again on a knife edge created by its overdependency on oil, and the manifest fragility of oil producers' ability to deliver their product to market affordably.

Holland Park, London, April 2011

Every so often you see people having what the peakists call their 'peak oil moment'. That's when they clock that there will be a big problem with oil depletion. I watched an Australian TV journalist go through hers five years ago, when she did a short film on peak oil for ABC TV. Now she has filmed an update, and I am watching it for the first time.[7]

Jonica Newby is a science reporter for ABC TV's flagship science programme *Catalyst*. She has been to the IEA in Paris to interview Chief Economist Fatih Birol. I can't wait to see this bit. Jonica I know to be a very persuasive woman. She is, not to put too fine a point on it, a charming and beautiful blonde.

I wonder how far she can coax Birol into going. In her original film she nudged an Australian oil company CEO into confessing that he saw a big problem with global oil depletion.

Her beginning sets a global scene. In this special *Catalyst* investigation, she says to camera, we travel from Paris, to London, to the outer-space-like world that is deep sea drilling to find out why so many industry insiders now say we'll soon look back on 2011 as the good old days when fuel was cheap.

Five years ago, when I first reported on this, the idea of world peak oil soon was sort of laughed at by the mainstream. So I'm curious – what are people saying now? Which is why my first stop is here.

Jonica on a street in Paris.

This is the headquarters of the International Energy Agency. It puts together the forecasts on oil production over the next 20 years – and it's the body many governments look to for accurate advice.

This multi-government agency is the voice of the mainstream. And I'm here to meet the man at the top, who just five years ago was confidently saying oil production will rise to 120 million barrels a day by 2030. But now?

Fatih Birol appears on screen.

When we look at the oil markets, he says, the news is not very bright. We think that the crude oil production has already peaked in 2006.

Jonica's voiceover, with the graph showing the IEA's 2010 projection of future global oil production now on the screen.

Hang on – did you get that? Crude oil production for the world peaked in 2006?

Birol continues. The existing fields are declining sharply in the North Sea, in the United States, in the Gulf of Mexico. Just to stay where we are today we have to find four new Saudi Arabias; this is a tall order.

OK, says Jonica. So what are their current projections, which include tar sands, natural gas liquids, all additional sources of oil?

We see the total oil production can increase up to 96 million barrels per day in 2035, Birol says. But this is potential. Nobody can guarantee me that the oil under the ground, especially in some key Middle East producers, will be developed and will be brought to the markets in a timely manner.

He holds his eyebrows high, and sits still, looking to the right of camera, at the interviewer, eyes steady. He has been media-trained. You can't read discomfort in his body language. He must surely feel it though.

So – no overall peak oil in sight officially, Jonica says. But not the reassurance I was expecting, and, frankly, it sounds like he's not confident in his own graphs. Which is interesting, because a team of Swedish scientists say they've proven the IEA projections have a fatal flaw.

Her film moves on to Sweden, and the Association for the Study of Peak Oil team at the University of Uppsala researching why Birol is so unsure of his graphs.

The IEA is making unfeasibly high assumptions about future extraction rates, Professor Kjell Aleklett explains.

She jumps to the Gulf of Mexico to look at the difficulties of deep-water drilling.

We've got this rush, says an industry insider on camera, this push to develop these very hazardous reservoirs, and the industry is starting to scare even me.

Yesterday, I reflect, with tar still washing up on the beaches, BP said it expected to be drilling the Gulf again within months. Of nearly five million barrels spilt, over a million barrels still cannot be accounted for by scientists studying the spill. Nobody knows yet how bad the long-term impacts will be. We may not know for a while. BP is trying to find ways to control scientific research on the Gulf spill, e-mails have shown. The company is making scientists studying the spill sign three-year confidentiality contracts.

Jonica goes to London to hear why big businesses outside the oil sector are worried. I watch myself explain.

The film moves seamlessly to a conclusion.

For final comment, she says, I'd like to return to that former bastion of 'she'll be right, we've got plenty oil' – the International Energy Agency, whose graphs officially show a gentle rise in oil supply over the next 20 years.

This must be the bit where she pressed the IEA's Chief Economist to the max, using her best Bondi Beach smile.

The time is running out, Birol says. The oil is today our lifeline, it is everywhere in the economy; if the prices go up or if there's a supply disruption this will be definitely very bad news.

How urgent is this?

I think, Birol says, it would have been better if the governments had started to work on it at least ten years ago.

Ten years ago? I erupt at the screen, can't stop myself.

My wife looks at me disapprovingly. She is Japanese and admires people with calm attitudes.

For most of those you have belittled this problem, I bluster, sat squarely with the oil industry calling the peakists scaremongers. Your staff still do, for God's sake.

Houston, says Jonica, it's just possible we have a problem.

Chapter 13

The anti-Oil Shock Response Plan plan

The financial crisis continued unfolding in early 2011 in a way that promised trouble even if the oil markets were able to deliver affordable oil. In early March, Bank of England Governor Mervyn King warned that the markets were at risk of another financial crisis, unless banks were reformed. Many warning-watchers were by now becoming exasperated. 'Weren't regulators supposed to act as well as warn?', I heard many times at this point. Kingfisher CEO Ian Cheshire, a respected captain of the FTSE 100, was particularly blunt. 'We need a radical reappraisal of capitalism', he said.[1]

In late March, ratings agency Moody's warned that banks might start taking outsize risks in order to win advisory business in an effort to gamble their way out of trouble. In the first week of April, Bob Diamond proved the point. 'Barclays must increase its risk appetite', he announced. His gall seemed to know no bounds. He was endeavouring to build the biggest investment bank in the world on the back of a government guarantee. Whether he knew he was doing this on the back of the yet-to-break Libor fraud, and the potentially criminal activities of his bank's emissaries in the Qatar bailout deal, future courts will decide. Whatever, he wanted a bigger risk appetite, and he was leading from the front.

The disasters for national economies that had been triggered by the risk appetites of investment bankers continued to unfold. Moody's cut Portugal's credit rating, following Fitch and S&P. Bailout loomed for the beleaguered country.

Despite everything, the UK's Banking Commission report failed to call for the break-up of UK banks when it reported in mid-May.

Books on the financial crisis were by now thick on library shelves. William Cohn published a particularly interesting tome at this time alleging that Goldman Sachs was playing dumb in order to get off the fraud hook. The bank had clearly shorted mortgage-backed securities at the same time as selling them, then obfuscated in congressional investigations, he said. *Rolling Stone* magazine also pulled no punches on 'the vampire squid', as the once revered investment bank was becoming known. Describing how a Senate sub-committee had laid out the evidence, Matt Taibii concluded that Goldman Sachs should now stand trial. 'If the evidence in the Levin report is ignored, then Goldman will have achieved a kind of corrupt-enterprise nirvana. Caught, but still free: above the law.'[2]

***

As for the risk takers in the energy world, the advocates of the shale gas panacea seemed to be enjoying mixed fortunes. Their institutional cheerleaders did a good job of putting out eye-catching numbers, mouth-wateringly so for those unconcerned about climate change. In April 2011, the US EIA calculated that shale gas held the potential to add 40% to global gas recoverable resources, including areas outside the US. As though to begin proving the point, China drilled its first fracked well, openly seeking to emulate the US boom.

But at the same time worrying case histories were emerging about the health and environmental impacts of gas fracking even in Texas. Almost every county in Pennsylvania had its open stories of the dangers of gas drilling in the Marcellus Shale. And research at Cornell University suggested that fracked gas has a worse carbon footprint than coal over a 20-year time frame, if the entire value chain is considered.

Industry advocates contested all these developments, of course, but they clearly had a potential setback on their hands in what they called 'social licence'.

In France, the government was openly pondering a national ban on shale gas exploration. Of course, they had a nuclear industry to protect.

The UK, meanwhile, remained open for business to gas and nuclear both.

London, May 2011

Finally the UK Industry Taskforce on Peak Oil and Energy Security is granted an audience with the Secretary of State for Energy and Climate Change, Chris Huhne. All the companies are represented. Huhne is accompanied by a large team of ministry officials.

John Miles of Arup, our chairman, opens the proceedings with a summary of our concerns. He has been brilliant through the preparation and production of both Taskforce reports. He likes to say: we are not a campaigning organisation. When he does, I sit quiet, thinking: John, if this isn't a campaign, I don't know what is.

But you are talking to an economist, Huhne responds, and, as an economist, I recall that the two previous oil shocks ended in recession, prices fell, and everything settled down again. What's to worry about?

Miles calmly observes that the third will be different, and explains why. Then he adds something worthy of Newsnight's Paxman.

And anyway, Secretary of State, surely you're not suggesting global recession should be the policy response to an oil crunch?

Huhne goes into reverse gear in a striking way. This is a clever man. I am surprised with his intellectual sloppiness. Maybe he has something else on his mind, I think. He is being questioned by police at the moment about a speeding ticket many years ago. They suspect he asked his wife to pretend she was driving the car, and take his penalty points. That would be perverting the course of justice.

So what you want, Huhne says, if I understand it right, is for government and industry to work together on an Oil Shock Response Plan.

We have never called it that ourselves, but that is precisely what we want.

Exactly, says SSE CEO Ian Marchant. We need to be asking ourselves what we would do now if we knew oil were $250 in 2014. The plan must major on transport. We may have to accelerate electric vehicles, for example.

Certainly I see no harm in 'no regrets' policies, Huhne says. Maybe there should be an emergency plan. You are talking about five years to the peak, everyone else says we have 40 years. It makes a lot of sense. We need a conversation about risk, probability and response.

Chris Barton, head of international energy security in the ministry, appears agitated at this point.

We have had conversations about all this, Secretary of State, he says.

He turns to me.

I can't understand what else it is you want us to do that we are not doing.

Chris, we have had hurried conversations in corridors at conferences, I say. We're after something much more serious than that.

We've been clear in two reports, John Miles adds. We want to know how we can help.

I'm absolutely open to this, Chris Huhne says. The Treasury's view is don't scare the horses. You need to talk to the Chancellor. And write to me with some suggestions for the process of drawing up an Oil Shock Response Plan.

We will, says John Miles.

As we prepare to leave, Ian Marchant makes a mischievous reference to speeding. I wince, but Huhne responds cool as a cucumber.

I'm in favour of raising the speed limit for electric cars, he says.

There is admiring laughter. Many of us assume he is the victim of yet another Metropolitan Police cock up.

The meeting ends. We spill out onto the pavement, self-congratulation in the air, pondering what to do next. We need to put out a press release, someone says. Perhaps we should check with the ministry first, another suggests. No need, says a third, the outcome couldn't have been clearer.

The company press offices confer. It is no easy matter to get the press offices of five companies to agree on something of this significance.

The press release goes out on 23 May. Its title: Government to work with business on plans to tackle peak-oil threat.

I am now in Japan. I watch the press for the next few days bewildered as to why no articles are appearing. When the IEA warns of an oil shock, much less a national plan for addressing one, it makes front-page headlines.

On 28 May, the reason becomes clear. The DECC press office must have been pouring cold water on the story. Chris Barton sends me an e-mail. We never agreed to such a thing, it says. All we agreed to do was look at the recommendations in the Taskforce's next report.

I suddenly feel like I am living in an episode of *Yes Minister*, the BBC TV comedy about Whitehall.

You must have been in a different meeting to me, I respond. Mr Huhne invited our chairman to write suggesting how the Oil Shock Response Plan should be drawn up and that's exactly what he is in the process of doing.

We have to wait fully six weeks for a response to John Miles's letter to Chris Huhne. It is Chris Barton who responds, though Huhne signs the missive.

As I said in the meeting, it reads, I would welcome your suggestions on measures which we might consider to reduce or mitigate the risks, additional to those already in train, particularly focused on low/no regrets measures. We discussed that this might be done by means of updating your reports on the oil supply challenge and including more detail on the specific measures you would like Government to consider.

I can only imagine the conversation that Chris Barton and the other officials must have had with their minister after we left that office.

In April 2011, French oil giant Total bought into solar energy in a big way. It acquired 60% of American PV manufacturer Sunpower, which was in need of heavy-hitting assistance in competing with low-cost Chinese manufacturers. This $1.38 billion deal was a first. I had been expecting conventional energy giants to be doing this kind of thing before this, as a hedge against unaffordable oil prices and climate policymaking. All the other oil giants seemed set on going the other way, however.

Total was also keen on nuclear, as I knew all too well from my debate in Norway with their CEO. How were their investments in that sector looking, at this point, compared to their solar play?

By mid-May 2011, it was clear that more than one of the Fukushima reactors had melted down. Prime Minister Naoto Kan told the Japanese parliament that the plant was still six to nine months from cold shutdown. The post-mortem was painting an increasingly shocking picture, and by now blame was being cast well beyond the operators of the Fukushima plant. Lawsuits trying to stop nuclear plant development elsewhere in seismically active Japan had posited just the sequence of events that had happened in Fukushima. But nobody had listened to the ageing protesters who had brought them to court. These people were now being feted as wise truth tellers. Industry officials who had demonstrably ignored or concealed dangers were being widely demonised. A culture of nuclear collusion was becoming clear in Japanese society, involving the government, nuclear regulators, plant operators, and even the courts.

In France, EDF, the main nuclear utility, and Areva, the main reactor builder, both majority-owned by the state, carried on as though there were no lessons to learn. They announced that they would be limiting maintenance in their twelve seaside reactors during the summer. They needed to keep them operating because the most intense drought for half a century was meaning insufficient water in the rivers used for cooling of their inland reactors, the majority of their fleet of 58. In May 2010, a third of French nuclear plants had to shut in a heatwave, and electricity had to be imported from the UK.

And this is the technology that is supposed to save us from climate change, the wags were quick to point out. A little heat and drought, and most of it can't even operate in its number one champion nation.

Switzerland was having none of this, and in late May they elected to join the Germans in phasing out nuclear power. Their five ageing reactors were providing 40% of Swiss electricity.

In June, Japan admitted the fuel in three Fukushima reactors was likely to have melted through their pressure vessels. This was a wholly more serious situation than even core meltdown. Italy voted overwhelmingly against nuclear power not long after this announcement. In Germany, Eon now accepted the inevitable. Just weeks after announcing a lawsuit against Chancellor Merkel for stranding their nuclear assets, CEO Johannes Teyssen now said: 'Germany will become a laboratory for the accelerated switch to renewable energy. Eon will position itself in this process – and then take what it learns out into the world.'[3] The utility had turned full circle on its earlier insistence that renewables be suppressed in the UK so as not to crowd out nuclear.

In France, President Sarkozy committed another billion euros to nuclear, saying there was no choice. He described post-Fukushima fears over nuclear safety as 'medieval'.

A particularly damaging aspect of the post-mortem under way globally was the issue of insurance of the world's 443 nuclear power plants. Almost all of them were uninsured, including Fukushima. Rather like the investment banks, they were relying on governments and hence taxpayers to bail them out if something went wrong.

For me, a particularly worrying theme was the nuclear industry's self-defence process. I already knew that they were working in deadly below-the-radar ways to undermine renewables. It was worrying to have it confirmed by no less a figure than a former French government minister, however. In June, Corinne LePage, a lawyer who had been environment minister, was interviewed by *Libération*. 'The nuclear industry sabotages renewable energies, I'm convinced of it', she said. 'Every time an industry was getting mature, it has been killed. It started with on-shore wind energy, then PV. Now we're being told that the universal industrial cure-all is off-shore wind energy. I'm willing to bet that it will also be demolished.'[4]

***

By June 2011, the International Air Transport Association was warning that the high oil price, $115 a barrel at the time, was threatening the viability of many of the world's airlines. Mighty Qantas was offering all its staff redundancy, hoping that exactly the right fraction would take them up on the offer so that they could cut enough costs to stay in business.

If they were in this much trouble at $100 oil, how would they fare at $200 oil or whatever lies beyond peak panic over peak production?

The annual publication of BP's Statistical Review of World Energy failed to do the job of easing concern. 'Running dry', read a headline in the *Economist*. 'Oil production fails to keep up with demand.' In 2010, world consumption had exceeded supply by five million barrels a day for the first time ever, the magazine noted. World stockpiles were being run down, it concluded.

The *Economist* is not a magazine noted for a tendency to panic, or a shrill tone of voice. It seems extraordinary to read these words, looking back from industry rhetoric about a 'New Era of Fossil Fuels' at the time of writing, less than two years on.

In its monthly report, OPEC estimated that world demand for its oil would average 1.7 million barrels a day more than the member nations had produced in May. This was enough to meet demand in an economy the size of France.

The IEA duly urged OPEC to lift their production, or else threaten the global recovery. Otherwise the 1.6 billion barrel stockpile in the 28 IEA states would need to be tapped for the first time since 2005. This too seems stupendous, looking back from the IEA's cheerleading about America's ability to be self-sufficient in oil in 2013.

The chorus advancing shale gas and tight oil as a substitute for crude oil had yet to find its full voice. China put out a tender for fracking under which only Chinese companies could bid. Shale gas deposits were believed to be abundant in China, the *FT* reported, though no gas had been produced yet. In Poland, a nation desperate for the shale gas dream to come true so that they could escape Russian gas dependency, expectations were running high for Exxon and Chevron's exploration programmes.

But, back in Texas, drought was threatening the gas and oil boom. Officials were limiting water use in the face of the worst drought since records began 116 years ago. A single fracked well can require as much as 13 million gallons of water over its lifetime, enough to supply the cooking, washing and drinking needs of 240 adults for an entire year.[5] In the UK, fracking was suspended until further notice after a minor earth tremor in the very first exploration well. As an *FT* headline put it, 'the golden age of gas may be a call too soon.'

In the USA, people were emerging at this point who had grave doubts about the shale gas bubble. On 25 June 2011, the *New York Times* ran an article detailing their concerns. After analysing hundreds of e-mails between relevant players, the *Times* reported that 'energy executives, industry lawyers, state geologists and market analysts voice skepticism about lofty forecasts and question whether companies are intentionally, and even illegally, overstating the productivity of their wells and the size of their reserves. Many of these e-mails also suggest a view that is in stark contrast to more bullish public comments made by the industry, in much the same way that insiders have raised doubts about previous financial bubbles.'

In one e-mail, an analyst for IHS Drilling Data wrote: 'the word in the world of independents is that the shale plays are just giant Ponzi schemes and the economics just do not work'.[6]

The concern here begins with the unexpectedly high drop-off rate in production generally encountered by the industry, relative to conventional gas wells, as drilling an individual shale gas well progresses. This means that if production levels in a particular shale field are to be maintained, an increasing number of wells have to be drilled. Because the easy wells tend to be drilled first, the later wells become more numerous and/or technically challenging and hence more expensive. Huge sums of money have to be borrowed to keep the operation rolling. This is an easy task given the perceived wisdom pushed in Wall Street that shale gas is the next hot investment prospect. What will be less easy will be the later task of servicing the debt, given that so much drilling is going on that the gas price is low, meaning low cash for interest repayments. Though the gas price may rise over time as drillers pull out of the game, a low gas price is a core rationale for the whole fracking enterprise, perhaps especially in the eyes of those overseas who seek to emulate it.

On 28 June, the New Jersey legislature voted for a ban on fracking in the state. On 29 June, France banned fracking outright. Environmental concerns about the process were beginning to sway politicians.

Neither should natural gas be seen as a climate-change panacea, the IEA warned. Reliance on gas to the extent the industry envisages would inflict a 3.5°C increase in global average temperature on the world, and risk irreversible global warming.

IEA Executive Director Nobuo Tanaka and Chief Economist Fatih Birol also warned that renewables could be muscled out in a global dash for gas.

That was considerate of them. The renewables industries were feeling a certain muscularity in their dealings with the energy incumbency at this time. My own company was encountering EDF's vicious pushback against solar energy in France. The company, effectively a ministry of the French state, had ensured that a workable solar feed-in tariff was cut to the bone in June. Solarcentury was in consequence *en route* to closing down its once successful French office completely.

This muscularity was due to worsen.

***

One view of the increasing incumbency pushback around this time was that investment in renewables, despite all the difficulties, was beginning to show real promise of breaking some clean-energy technologies through to tipping points. A record $211 billion had been invested in clean energy in 2010, up 32% on 2009, with $50bn of it in China. The main drivers were wind in China and solar rooftops in Europe, $34bn having been invested in German rooftops alone.

Where could this trend lead, given a fair wind? In May 2011, the Intergovernmental Panel on Climate Change produced its first report on renewable energy, a 1,000 pager on which many of the world's renewable-energy experts had collaborated under the UN's leadership. Eighty per cent of global energy could come from renewables by 2050, the IPCC concluded, provided governments maintained their market-support mechanisms. Crucially, renewable energy deployed at this scale could keep greenhouse-gas concentrations below 450 parts per million. The process would be affordable: replacing conventional energy on this scale would cost only some 1% of GNP per year. And progress to date, though far from perfect, was not bad: of 300 gigawatts of new electricity capacity installed globally in the years 2008 and 2009, 140 had been renewable.

In the UK, the government's Climate Change Committee echoed a small degree of this optimism, concluding that 45% of UK energy could come from renewables by 2030. In Scotland, the national government committed itself to the extraordinary target of 100% of electricity supply by 2020. First Minister Alex Salmond, making his announcement while opening an online portal showcasing offshore wind and marine technology, was evidently bidding to secure Scotland's place as the green-energy powerhouse of Europe.

Two things struck me about this. First, most of the UK's remaining oil and gas lay in Scottish waters. Though they were fast depleting, there would still be plenty left by 2020. Second, relatedly, if Salmond were to achieve his remarkable ambition, he wouldn't be able to rely on the energy incumbency to help him: they were either opposed to major renewables deployments or slow to implement them. His government would need to spawn hundreds of community energy projects very quickly.

Wadebridge, UK, July 2011

Around the nation, the Transition movement is flourishing. As people start acting in groups, no longer waiting for government, community renewable energy projects are becoming commonplace. Wadebridge, a town of 10,000 in Cornwall, has decided it is going 100% renewable. The Wadebridge Renewable Energy Network (WREN) has 50 projects up and running. Solarcentury is collaborating with them. They have invited me to come down for a day and inspect what is going on.

Stephen Frankel, the founder, sits in a friend's kitchen explaining to me how WREN managed to get 600 people into the small town hall when they announced their formation, and why his group is opening an energy shop on the High Street.

It's that idea of autonomy and protection from personal economic threat that does it, he says.

We go to visit a farmer who has installed a 200 kilowatt solar array in one of his fields. It is WREN's most ambitious project yet. Some of the profits from the project will go to fund the organisation. It is official opening day. I tour the site with a few dozen local people. The panels stand proud in rows across a field near the farmhouse, inclined to the sun. Between them, a flock of sheep grazes.

And people say solar uses up too much land that could be used for agriculture.

Solarcentury's marketing folks, ever vigilant for opportunities, have sent a cameraman with me. They want to make a short film or two about what is going on. [7] Harriet Wilde, a young spokesperson for WREN, gives her view of it all for the camera. Behind her as she speaks, gorgeous countryside rolls away to a far off estuary.

I think we've got a fantastic community in north Cornwall where we're all very involved, she says. We work together as a community in all aspects. So renewable energy is just the next step. We're making money as a community. We're dividing it up as a community. It's a win-win situation. [8]

I am beginning to be in particular need of days like this.

Chapter 14

A 'bollocks' subject

The get-rich-on-coal trend that so worried climate activists about the Coal India IPO continued into 2011. Mongolia's vast national coal reserves were the next big prospect for flotation. Every investment bank sent its champions to Ulan Bator to fight for a piece of this gold rush. Some were so keen, the *Financial Times* reported, that they ended up brawling in a bar. Elsewhere in the capital markets, commodities trading giant Glencore was positioning to raise multiple billions in an IPO before moving on to a merger with mining giant Xstrata. Coal was written all over these deals, and 'risk' posed by climate regulators living up to their responsibilities and promises would not be featuring high on the list of issues drawn to shareholders' attention. In Glencore's 1,637-page prospectus for its initial public offering, only a single paragraph noted climate risk.

This blindness on the capital markets is a structural problem. Back in 2007, a group of concerned financial analysts led by Mark Campanale, then working at Henderson Global Investors, had coined the term 'unburnable carbon' to describe the share of fossil fuel reserves that could not be used in a low-carbon world. They had been trying unsuccessfully to find funding to quantify the extent of the problem. Good ideas sometimes need opportune moments – which came paradoxically with the failure at the Copenhagen climate summit. Ruing the failure of world leaders at that summit, I wrote an article in the *Guardian* warning of the unfettered financing of carbon fuels that could now be expected in the capital markets. Joanna Messing, adviser to an American foundation, the Growald Foundation, contacted me to ask if I knew of any work by financial analysts warning about the risk, and quantifying it. I was able to hook her and the Growald family up with Campanale, and so the think-tank Carbon Tracker was born. The Rockefeller Brothers Fund joined the Growalds in providing start-up funding. Mark asked me to chair it, and I jumped at the chance.

I and other scientist environmentalists had been warning about what we thought of as 'the carbon arithmetic' since 1990: the fact that there is way more carbon in fossil-fuel reserves than can be burned if the world is serious about staying below 2°C of global warming. Now the task was to put numbers on the excess right down to the company and stock exchange level. Climate regulations had been tightening around the world, but not sufficiently to stop global emissions rising or prevent the capital markets from becoming more fossil-fuel intense. The major coal, oil and gas focused companies, with the support of the investment banks and their investor clients, continued to be the most successful raisers of capital on the world's leading stock exchanges. It had to stop.

The Carbon Tracker idea for slowing the dysfunctional process was to force the markets to recognise that a bubble was building. This we intended to do by persuading key players across the financial chain to recognise the risk of their supposed assets being devalued.

The carbon dioxide potential in the world's proven fossil-fuel stocks was some 2,795 gigatons in 2011 – five times more than climate scientists were saying at the time that we could safely burn for the planet to have an 80% chance of keeping to a 2°C warming target. If emerging climate policy-making were to begin instigating measures aiming for that target, not all of the carbon-fuel companies would be able to develop their assets. This risk goes completely unrecognised by all sectors of the financial chain as things stand. In the event that a critical mass of policymakers starts to become serious about emissions reductions, perhaps motivated by electorates increasingly scared by the emerging impacts of climate change, substantial assets would be stranded, to the detriment of value in pensions and other investments everywhere. The sudden realisation of that in the markets would amount to another bubble bursting, and a grave shock to the global financial system.

Here is how the logic of the carbon bubble works. Scientists from European centres of excellence in climate science calculated in 2009 that if 886 billion tonnes of CO_2 is released globally during the period 2000–2050, there is a 20% chance that global warming will exceed a 2°C hike of the global average temperature above pre-industrial levels.[1] By 2011 we had already burnt over one-third of this 886 billion tonnes CO_2 budget. The known fossil-fuel reserves, catalogued each year in the BP Statistical Review of World Energy, far exceed the remaining allowance of 565 billion tonnes of CO_2. The 80% of reserves beyond this limit are what Carbon Tracker refers to as 'unburnable carbon'.

To repeat myself, unburnable as in stranded, should climate change become an issue for action rather than endless stalling.

London, July and August 2011

KPMG hosts the launch of the first Carbon Tracker report, at its headquarters in Canary Wharf. Vincent Neate, partner for climate change, sets the scene. He confides that the absence of his energy colleagues in the audience is a symptom of the problem those of us worried about climate change will have in getting the climate message out to contemporaries operating in and around the energy incumbency.

Carbon Tracker has hired a brilliant analyst, James Leaton, ex-PWC, to lead in quantifying the disposition of unburnable carbon company by company and stock exchange by stock exchange. It is, surprisingly, the first time this exercise has ever been conducted. James summarises the results.

I chair a discussion after he has finished. So where are we most likely to first break the chain and get some sector or other of the financial world to start recognising risk, I ask?

We should focus our efforts on the investment banks, some say: they make the game.

Others think we should try using high-quality PR to alert people to this new form of bad practice in the financial sector. HSBC does no banking with military companies, or with unsequestered coal in developed countries, so can't others be persuaded to join the Co-op in eschewing fossil fuels altogether? Might this set reform going?

What strikes me most, listening to the discussion, is how the premise about the risk of stranded assets is not challenged.

Maybe the audience is self-selecting, I think. But no, there are some pretty conservative players among the City people present.

And so I watch the press commentary on the report afterwards with more than usual hope. John Elkington, long-term *éminence grise* of British sustainability thinkers, gets things off to a good start.

Having written my first report on climate change in 1978, he admits, I have been dutifully tracking the evolving science for half a lifetime, but only on Friday 15 July 2011 did I truly feel that the climate, carbon and financial agendas had been spot-welded in a way that potentially brings all of this right home to people such as asset owners, rating agencies, brokers, analysts, investment bankers, accountants, data providers and financial regulators.

Analysts and fund managers interviewed by the *Financial Times* about the arguments in the report do not appear to agree with him. One anonymous analyst is quoted as follows: 'I think it's a bollocks subject. I'm not interested in this kind of subject. I think this is complete hot air.'

I am encouraged, both that the *Financial Times* is prepared to print such a rude word — no doubt for the first time — and because the analyst's statement is a crude expression of a belief system, not a logical rebuttal of an argument. It tells the reader everything they need to know about the intellectual rigour at work here.

Says a fund manager: 'We apply our own judgement as to what is technically feasible, we don't look at what is legally feasible. There are no restrictions on companies at the moment in terms of the amount of oil and gas that they can produce.'

At the moment, Mr or Ms Fund Manager. At the moment. That is the whole point.

Then comes an article in the *Actuary*, the trade journal of those who endeavour to quantify risk in the face of uncertainty.

'In the Carbon Tracker work, maybe we have identified a new class of toxic assets', it says.

The peak-oil debate had long had a tribal air about it, but from the second half of 2011 it assumed a surreal element. Arch oil-industry apologist Daniel Yergin launched a withering assault on the 'peak-oil theorists' in a *Wall Street Journal* interview in September. 'There will be oil', the title of his article read. The CERA professor waxed triumphant. 'For decades, advocates of "peak oil" have been predicting a crisis in energy supplies. They've been wrong at every turn.' He predicted a plateau in production, beginning fully four decades in the future, with as much as 110 million barrels a day being produced by 2030. He added a caveat, of course: this would not be a done deal, because of geopolitics.

They were all playing that insurance card, I had long since observed.

His main arguments were essentially the same as we have encountered thus far in this book. The focus was on the total resource theoretically available, not on the practically deliverable flow rates. He also made much of the recent reversal of the US oil-production decline. The Bakken formation in North Dakota had gone from 10,000 barrels a day in 2003 to over 400,000 barrels, he exalted. North Dakota had become the fourth-largest oil-producing state in the country. Thanks to fracking, such tight oil could add as much as two million barrels a day to US oil production after 2020, something that would not have been in any forecast five years ago. Overall US oil production had increased more than 10% since 2008. Net oil imports had reached a high point of 60% in 2005, but today, thanks to increased production, biofuel use and greater energy efficiency, imports were down to 47%.

The peak-oil 'theorists' responded with their own battery of arguments, but this as ever managed only a fraction of the airplay that the comforting narrative did. One withering dissection, by British analyst Euan Mearns, particularly deserved a right of reply in the *Wall Street Journal*. Instead, it was to be found on the specialist website *The Oil Drum*. The cognoscenti would read it, but few others.

Mearns kicked off his attack by chastising Yergin for not talking about decline rates and their implications. Crude oil plus condensates plus natural gas liquids production had been running at around 82 million barrels a day since 2005, he said, but depletion was wiping out 4.1 million barrels a day of that capacity every single year. To get to 110 million barrels a day by 2030 would mean replacing that each year and then adding the rest, mostly from unconventional forms of oil. This was essentially the same point that the IEA had made back in 2008 when talking of the need to find six Saudi Arabias of extra production by 2030. And even then, if the whistleblowers cited earlier were to be believed, the figures had been massaged to make them less unsavoury to the Americans. If all the rest of that requirement could be added, said Mearns, how come the industry had only been able to keep production on a plateau for the last six years, even in a time of extended high oil prices? Yergin's two million barrels a day after 2020 had to be viewed in that perspective.[2]

Undaunted by this kind of reasoning, in October bullish investment-bank analysts predicted that US oil production could exceed that of Saudi Arabia and Russia within a decade.

In the UK, the same pattern of exuberant optimism was emerging. Fracking pioneer Caudrilla Resources announced that it had discovered 'huge' gas reserves under Lancashire. Based on just two wells, they professed their discovery would be the biggest single reserve in the world, equivalent to 20% of the whole of China's shale gas resource, supposedly the biggest globally.

Assuming their 'asset' assessment was correct, it would be interesting to hear what the locals would have to say about the access requirements. Caudrilla would need to drill some six to eight holes per square mile, potentially running to hundreds or thousands of holes eventually.

With their very first, they had triggered an earth tremor, enough to shut down UK shale gas drilling pending a full safety review.

In the US, environmentalists were making steady headway on restricting access by the industry. Fears of groundwater contamination had now led to bans in New York as well as New Jersey, and the EPA had outlawed fracking with diesel, a component in wide use by the industry, in the water pumped underground.

Meanwhile, conventional oil had now completed six years of essentially flat production. EIA data showed that production of crude oil had stopped growing in 2005 and then fluctuated between 73 million and 74 million barrels per day annually through 2010. Production had averaged 73.8 million barrels per day from January to July 2011. More than 14 million barrels a day were being added from gas production liquids, unconventional oil (mostly tar sands) and biofuels, so total liquids production stood at around 87 million barrels a day.

What would it take to keep this show on the road?

One important factor emerged at this time. The oil and gas industry would have to be prepared to work with criminals and thugs, evidently. For that was what papers submitted in court proceedings showed that BP considered their billionaire Russian partners in TNK-BP to be.[3]

***

In determining the fate of nuclear power after Fukushima, much would be dependent on how smoothly EDF and Areva, French global standard bearers for the industry, managed completion of Areva's two next-generation EPR nuclear reactors: at Flamanville in France, built for EDF, and at Olkiluoto in Finland, built for a Finnish utility. Even before Fukushima, things had not been going well. In July 2010, EDF had delayed the start of the 1,600 megawatt Flamanville reactor from 2012 to 2014 and raised its cost estimate from €4 to 5 billion. In July 2011, it announced that the reactor would be delayed for a further two years, until 2016, and cost €6 billion. The reactor would be requiring twice the time originally announced and nearly twice the budget.

EDF blamed the complexity of the engineering and the additional caution required in the wake of Fukushima. No reactor had been built in France for 15 years, and it seemed almost as though the industry had forgotten how to build one. Even design seemed to be an issue. In August, French nuclear regulators told EDF that they had thirteen areas of concern about the Flamanville reactor. Fixing the problems would require 'great efforts' by the company, the regulator's letter said.

EDF was hoping to lead the 'nuclear renaissance' in the UK by building two of these same EPR reactors at Hinckley Point, and two at Sizewell.

In September, the International Atomic Energy Agency (IAEA) announced that the Fukushima disaster was now as bad as Chernobyl. Six countries had by this time notified the IAEA that they had abandoned their nuclear plans as a result of the disaster. The UK looked like being the largest Western market for nuclear in the decade ahead, and certainly the most significant in terms of the long-term prospects for the nuclear industry.

**

Entering August 2011, two-thirds of Japanese nuclear power plants were offline, with only 17 of the 54 nationwide operating. Yet the lights were still on in Tokyo. A Herculean energy-efficiency drive was under way, and Japan was not particularly missing that fraction of the 30% of its electricity nuclear used to provide.

Before the disaster, if anyone had said that a modern nation like Japan could keep the lights on after the sudden withdrawal of 37 nuclear reactors, they would have been scoffed at.

Of course, Japan was now even more highly dependent on expensive fossil-fuel imports. It had a bewilderingly low 1% of electricity from renewables. From here on, a major drive would begin to accelerate deployment of these technologies. Japan was viewing Germany with envy now. With more than 20% renewables in the electricity mix, rising rapidly towards a target of 35% by 2020, a new energy world was taking shape there.

A common argument of nuclear advocates is that nuclear is essential for running the railways. Renewables won't be able to do that, they tend to say.

In August 2011, Deutsche Bahn announced that it had targets and time-tables for running the German railways 100% on renewables. It aimed to raise the percentage of wind, hydro and solar energy to power its trains from 20% now to 28% in 2014, *en route* to becoming carbon-free by 2050. Some local railways in Hamburg and Saarland were already running 100% on renewable energy.

Deutsche Bahn's trains transport 1.9 billion passengers and 415 million tonnes of freight each year at speeds of up to 300 kph (186 mph), using 2% of all German electricity: 12 terawatt hours, as much as Berlin with its 3.2 million residents consumes.

But despite such reasons for bullishness, the renewables industries were not faring well on the stock exchanges. In December 2009, the values of renewable-energy stocks began a steady fall. The failure of the Copenhagen Climate Summit that same month was clearly a major factor. A key driver for growth of renewables markets had been undermined. Government support for renewables, in the form of feed-in tariffs and other subsidies, had wobbled in the wake of both the failure of the climate negotiations and the new financial constraints after the crash. As I knew well by this time, intense incumbency lobbying − for nuclear and gas both − had taken maximum advantage of this. Germany would install fully 7.5 gigawatts of PV in 2011. The global market had grown tenfold in four years from 2007, from 2.7 gigawatts to 28.7 gigawatts, in the teeth of the worst recession in a century. Yet in October 2011, one investor, Altium Securities, went so far as to tell the *Financial Times* that solar energy had become 'uninvestable'.

The *FT* summarised the state of play in a headline: 'Patience is a virtue in hunt for game-changing "green google"'. The value of publicly traded cleantech companies had gone down from $475bn in 2007 to $142bn. Venture capitalists were taking flight from early stage funding. The question asked at the retreat in Utah was being answered. Investments in energy storage were outpacing solar for the first time.

The photo in the *FT* article was of Solarcentury roof slates. The main quote came from Solarcentury investor Vantage Point. My company and its Silicon Valley investor had become poster children for the widely perceived failure of clean energy as an investment prospect.

No matter how many German trains were running, and would be run, on solar and wind.

Forest Row, UK, November 2011

An energy fair, organised by the rural citizens of Sussex. An opportunity to debate with their MP, who happens to be the energy minister. Charles Hendry and I sit facing a few dozen of his constituents in a church hall, a Christian cross and a child's model house on a shelf behind us.

This is the Tory heartland, but still I have the feeling that Hendry is exercising a degree of courage coming here. This is a Transition Town event, and the government has just cut the solar feed-in tariff much more deeply than expected, and too deeply to make solar work economically.

Hendry does have to face a lot of hostility from his constituents over this and other elements of the government's green record. Clearly unused to this kind of reception, he is very red of face by the end.

In closing the debate, he makes his bottom-line point, and it is a revealing one.

If the big energy companies were to walk away from the UK, he says, as they could do, it would be a complete disaster.[4]

Facing the danger of contagion from Greece, eurozone leaders agreed a further bailout in June 2011. They tabled €109 billion and – in what amounted to a partial default – also called in private bondholders to contribute a further €37 billion.

In August, Standard and Poor's downgraded the USA's AAA credit rating. Fears of turmoil ensued. China reacted with an official statement: 'China, the largest creditor of the world's sole superpower, has every right to demand the United States address its structural debt problems and ensure the safety of China's dollar assets.' Beijing insisted the US should slash its 'gigantic military expenditure and bloated social welfare costs', and demanded – not for the first time – a new global reserve currency to replace the dollar.

In September, Germany almost doubled its guarantees to €211 billion, so that the European Financial Stability Facility, as it was called, had €440 billion available to it. The threat of Greek default had receded for the moment, but critics wondered if it would be enough to stop the rot.

In October, the Bank of England launched a second round of quantitative easing, injecting £75 billion into the ailing economy to try to boost demand

and prevent inflation falling too low. Britain was in the grip of the world's worst ever financial crisis, said the Governor, Sir Mervyn King.

The financial markets were 'at their wit's end', as an *FT* editorial put it in the midst of all this. The paper rounded on politicians for creating risks that investors were ill-equipped to deal with. By dallying in the eurozone and through brinkmanship over prospective US default, they had amplified already worrying downturns on both sides of the Atlantic.

If politicians were hardly in their element, neither were the architects of their reluctant trial by market. The inquests into bankers' malfeasance continued to appear. One calculated that bankers had paid themselves more than £2 trillion in the last five years. How much of this staggering figure was on the basis of lawbreaking was still open to question. In August 2011, Goldman Sachs felt the need to hire a top criminal lawyer to defend itself against charges of deception, a man who had in the past defended Worldcom and Enron executives, men later convicted of gigantic corporate frauds.

Still the banks escaped major reform. In the UK, in the face of strong lobbying by banks and the CBI, the government had given up until – they said – after the next election. This was supposedly for fear of inhibiting the freeing up of lending. Cynics accused them of having too many banker friends and funders.

The frustration was increasingly intolerable for many people. In October, 700 anti-Wall Street protesters were arrested in New York. Their group, Occupy Wall Street, had erected a tented encampment, among other forms of protest. As protest spread to other American cities, President Obama felt able to toughen his rhetoric. The banks 'can't be competing on the basis of hidden fees, deceptive practices or derivative cocktails that nobody under-stands and that expose the entire economy to enormous risks', he said.

In mid-October, the pan-US protests spread offshoot occupations in London, Sydney, Frankfurt, Hong Kong, Toronto and other cities.

Initial reaction to the Occupy movement was hostile, with Republican politicians speaking of 'angry mobs'. But as the opinion polls came in, they rapidly had to change their tune: there was strong public support for the protesters' cause, even among right-leaning voters. Some corporate leaders chose to express empathy for the Occupy cause in public, including the CEOs of Wells Fargo and GM. At Harvard University, 70 students walked out of an economics class to join an Occupy protest in Boston, demand-ing major changes in what they were taught. It seemed they felt that less Adam Smith and more Keynes would be appropriate for the world they faced.[5]

In November, the G20 leaders gathered in Nice to try to agree a plan for distressed countries. All attention was on debt-loaded Italy, now forced to accept IMF monitoring of its austerity programme. An Italian default would be a whole new order of problem. One hope was that the IMF's resources for battling the crisis would be increased by $250 billion to $1 trillion. But the leaders baulked at this.

George Osborne returned to the UK amid tumbling markets, with the prospect of recession looming again, and admitted that the Treasury was in crisis planning for a eurozone collapse.

City of London, November 2011

In London, the Occupy protest is as near to the Stock Exchange as it can get: a tent camp outside St Paul's Cathedral. The organisers ask me to air my reservations about modern capitalism, speaking as a supposedly successful modern capitalist.

If you believe the rightist press, the protesters in this tent camp are drugged-up anarchists. Hypodermic needles are to be found on the litter-strewn streets. All nonsense, of course. I find well-organised people, with genuine concerns, mostly university educated. I speak to them in a marquee they have named the Tent City University.

They are more than justified in their concerns. Panic is in the air in the markets once again. Things are very clearly going to get a lot worse before there is any chance of them getting better.

I stand in my pin-stripe suit and look around the tent at the dozens of colourfully casual people sitting on the carpets and lounging on cushions. A video camera is running to capture the discussion on the web for those who can't be physically there. I have to decide how frank to be.

I'm just wondering, I say, how many of you are Metropolitan Police officers in disguise.

They laugh good naturedly.

But it's an odds-on certainty that some of them are. The papers have been full of revelations about undercover police officers used to spy on climate protesters. The police seem much less keen on investigating and arresting financiers or energy executives than they are those who protest against their excesses. Yet those excesses so recurrently spill over into criminality. When they do, the finger-pointing comes from whistleblowers within companies, not undercover police investigators.

I feel a great well of empathy inside me for the people in this tent. Their future, in a system bent on institutionalised self-defence, is stacked against them.

I know I have a tendency to be emotional. I know I can get a little carried away by that. Mostly I can keep it disguised. But today, I know, I am going to have to pick my words very carefully.

Chapter 15

To the point of being suicidal

Hopes were low ahead of the December 2011 annual climate summit in Durban. The 2010 summit, in Cancun, had managed to keep the negotiations afloat, ending with an accord by governments recognising that their goal should remain holding the increase in global average temperature below 2°C above pre-industrial levels, extending their ambition even to 1.5°C if emerging evidence of climate havoc required. The negotiations in Durban aimed to build on that agreement.

They came within just minutes of falling apart completely. But delegates managed to pull back from the brink. They agreed to start work on a new climate deal that would have 'legal force' and require both developed and developing countries to cut emissions consistent with the Cancun goal, terms to be agreed by 2015, at a climate summit in Paris, and to come into effect from 2020. It was something. It kept the process alive, albeit still on the road to a four-degree-hotter world, absent the strong actions needed to make the aspirations come true.

However, just a day after signing the Durban accord, Canada pulled out of the Kyoto Protocol. Their message was clear: it was now going to be tar sands or bust for them. A chorus of international outrage duly descended on Ottawa. China called their decision 'preposterous'.

Their stance put huge political pressure on President Obama. The oil industry was desperate for him to approve a pipeline from the giant tar sands deposits in Alberta down to the US south coast, the Keystone XL pipeline. The American Petroleum Institute made it very clear that if he didn't do this, he would face the full force of their chequebooks in the run up to the November 2012 election.

This set the stage for a pivotal confrontation. American climate campaigners knew that if they lost this battle, they – and the world with them – would be much less likely to win the carbon war. Veteran American climate campaigner Bill McKibben and the organisation he had founded, 350.org, joined with other environmentalists to form a high profile civil protest movement intent on stopping the pipeline.

***

Entering 2012, the oil price in euro equivalent was nearing its July 2008 peak. The average oil price in 2011 was a record $107, up 14% on the previous record year, 2008. It was heading for an average of $111 during 2012.

Another warning emerged in January about the IEA's ability to offer reliable advice on oil. Oliver Rech, an oil analyst who had developed scenarios for the agency between 2006 and 2009, gave an interview to *Le Monde*. He said that the prospects for oil production were now much more pessimistic than those published by the IEA, that production for combined totals of conventional and unconventional oil would certainly remain below 95 million barrels a day, and that there would be an inevitable overall decline of oil production somewhere between 2015 and 2020.[1]

By complete contrast, senior Citigroup banker Peter Orszag professed on 1 February that the fracking boom would extend from shale gas to tight oil, and so could 'finally cap the myth of peak oil'.[2]

The very same day, Exxon's first two wells drilled in a Polish shale formation failed to find commercial quantities of gas. So much had been hoped of these Polish deposits. An April 2011 assessment by the US Energy Department predicted they held enough gas to supply domestic consumption for more than 300 years.

On 17 February, undeterred, Citigroup told the *Wall Street Journal* that 'peak oil is dead'. After decades of decline, US oil production was now on the rise, entirely because of tight oil production. Tight oil could add almost 3.5 million barrels a day to US oil production between 2010 and 2022, they said. It had already slashed a million barrels a day from US oil imports. 'One day it may allow the U.S. and Canada to be self-sufficient in oil', Citigroup enthused.

It was clear to many following the detail that all this was an effort to talk the market up.

The investment banking community itself was becoming polarised on oil supply. In the real world, the outlook for global supply was looking grim, inventories were low and risks to output were spreading. Serious people were publicly doubting Saudi Arabia's ability to make up for any shortfall. A report by Deutsche Bank analysts Mark Lewis and Michael Hsueh majored on the soaring domestic demand within OPEC and other major exporters and warned that while global crude production had been stagnating at around 74 million barrels a day since 2005, global crude exports were on a downward trajectory.[3] Their conclusion was that there hadn't been such a serious threat to global oil supply since the late 1970s and early 1980s.

Endeavouring to demonstrate spare capacity, the Saudis opened the kingdom's oldest oilfield, Damman, for the first time in 30 years. In prospect from this mothballed field were 500 million barrels of oil that might yield as much as 100,000 barrels a day of heavy crude. Not much of a prize, and of the wrong species, in a world desperate for light, sweet crude.

If there is no peak-oil problem, peakists asked, then why is Saudi opening old wells for heavy crude?

On 25 February, the oil price hit $125, on fears over Iran's uranium build-up. It was now the highest it had ever been in sterling and euro terms. 'Soaring oil prices will dwarf the Greek drama', Liam Halligan predicted in the *Daily Telegraph*, speaking for growing numbers of commentators. 'Crude is now expensive not due to political argy-bargy but because of the fundamental truths of demand and supply.'

But on 6 March, reporting on the annual CERA conference, the *Wall Street Journal* offered this: 'These are happy times for the oil industry, which has shed talk of peak oil for the thrill of new discoveries.'

At the same conference, Total's upstream boss, Yves-Louis Darricarrere, volunteered that he expected peak oil to be just around the corner. The world would find it difficult to produce more than 95 to 97 million barrels per day in the foreseeable future, he suggested, not much above the 91 million barrels per day or so expected by the IEA during 2011. The problem, he explained, was in essence that although the shale story was impressive, the depletion rate in old crude fields, and the massive requirement to replace them, was even more impressive.

Two days later, Exxon CEO Rex Tillerson admitted that fracking was failing so far in China and Europe. Some shale formations in Europe and China were proving impervious to the drilling techniques that opened the vast reserves of natural gas and oil from Texas to Pennsylvania, he told analysts in New York. New methods and tools would need to be invented to tap many of the shale fields that energy companies and governments expected eventually to yield a bonanza of fuel.

Saudi Arabia said it had the capacity to raise production by 2.5 million barrels per day if markets needed it. The kingdom was already pumping at ten million barrels per day, its highest rate for 30 years. These were merely verbal assurances, sceptics were quick to point out. Where was the proof they could actually lift production that much?

On 15 March 2011, with warnings of renewed recession thick in the air, the US and UK actually floated the prospect of emergency releases of oil stocks from reserve. The margin for error was that tight.

There was by now a surreal element to the peak-oil risk debate, it seemed to me. It was polarising into two extremes. Logic seemed to have flown out of a window somewhere.

I wondered how many neuroscientists were following the pattern of events.

I was too busy to find out.

***

The nuclear industry, meanwhile, had no pretensions that it could help with the high price of oil any time soon, given the decade-plus it admitted it needed to build reactors. All it wanted was to be in the frame for electricity generation, in as many countries as would still allow it.

Whether one of these would be Japan seemed very much in doubt. Tepco succeeded in bringing the Fukushima plant to cold shutdown on 16 December 2011. The radiation leaks from the three reactor meltdowns had forced more than 100,000 people to abandon their homes and polluted some 3% of Japan's land mass to levels requiring decontamination. Opinion polls were showing strong anti-nuclear sentiment in a previously quiescent population. By February 2012, 52 of 54 Japanese nuclear power plants were offline. All 54 would be off by the spring, with many too old to be brought back on.

The absence of so much nuclear electricity was not proving to be anything like the problem pundits had warned.

And so the nuclear focus fell with even greater intensity on the markets where the industry could see a viable beachhead for re-establishing itself.

In the UK, EDF and other big energy firms were found in December 2011 to have embedded at least 50 full-time employees in government departments over the previous four years, free of charge and working as civil servants for up to two years. The renewables industries had none. The traffic was in information, as well as lobbyists disguised as officials. The same month, documents obtained under the Freedom of Information Act showed that DECC had been passing sensitive documents on government policy to the Nuclear Industries Association. It seemed that many top civil servants badly wanted the nuclear renaissance to stay alive, despite everything.

Such stacking of the energy cards led to the inevitable. The UK's electricity market reform bill ought to have been a document fit for the much changed energy circumstances, one that Silicon Valley, perhaps, might have approved of. Instead, the draft was a crib sheet from the nuclear and gas industries' wish-lists. British Gas, via its parent Centrica, was a partner in EDF's UK nuclear operations. EDF also had gas generation assets in the UK. On electricity market reform, therefore, they worked together to spin their solutions as cheaper in the long run than renewables, meanwhile undermining renewables.

The vast majority of the UK's electricity and gas market is held by just six companies: EDF, Centrica, Eon, RWE (owner of npower), Iberdrola (owner of Scottish Power) and SSE. These giants had long been known as the Big Six.

All six utilities were now under investigation by the regulator Ofgem because they had been mis-selling electricity and gas to consumers. In August 2011, the *FT* reported that the companies were so hated by the public, and electricity and gas bills were inflating so fast, that the political backlash threatened the very viability of their business models.

In writing about the Big Six, I have faced a dilemma. One of them, SSE, is a minority investor in my company, as I have noted earlier in this history. Their CEO, Ian Marchant, is a man I like and respect. He is a generous supporter of SolarAid and has been a key player in the UK Industry Taskforce on Peak Oil and Energy Security. I know he tries harder than any other Big Energy boss to promote the clean-energy agenda. SSE has installed more renewables than all the others and did not play the game of embedding lobbyists in government ministries. Marchant has urged me, along the way, to start referring to the Big Five, not the Big Six. He has a case, but when it comes to general corporate character, a quick google of 'SSE mis-selling' will show a sorry chronicle. They are in the same camp as the others on ticking the character traits deemed necessary by the investors who drive modern capitalism. The fact that this is true despite the best efforts of a good-hearted CEO tells us much about the nature of the system.

Even after the *FT*'s shocking conclusion about the threatened viability of their business models, the Big Six made few noticeable changes. Entering

2012, the bosses were pocketing up to £4 million a year while more than a quarter of UK households were in fuel poverty. (Ian Marchant, it should be added, was lowest on the pay list, and he donated a generous chunk of his salary to charities.) The energy firms received no less than four million complaints in 2011. Consumer Focus accused them of treating their customers with contempt.

Meanwhile, most were showing a clear preference for gas. Eon, RWE and Iberdrola had by mid-April all made noises about pulling out of their UK nuclear ventures. EDF was in danger of being last man standing when it came to nuclear power.

Most worrying for future energy policy, renewables investment in the UK was now suffering grievously in the face of Big Energy's attempted dash for gas, and no doubt the antics of the dozens of lobbyists implanted in Whitehall ministries. Less than a gigawatt of wind had been approved in 2011. Meanwhile 30 gigawatts of gas was at planning stage. As for solar, the government had attempted an ambush cut of the feed-in tariff that would have crippled the industry. Feed-in tariffs are designed to be cut in manageable increments, keeping the supported technology economic until such time as the subsidies are required no longer. A precipitate cut, sprung on an industry and so leaving it no time to adapt, would be a classic tactic for a wrecking agenda in officialdom.

Two companies, Solarcentury and Homesun, took the government to the High Court, alleging they had acted illegally in ambush-cutting the feed-in tariffs for solar.

In December 2011, we won the case.

***

Entering 2012, the eurozone crisis was escalating worryingly. As of November 2011, the government bond market across Europe had essentially ceased to function. Vital parts of the eurozone economy were cut off from credit. The central banks stepped in to cut the interest rate on emergency dollar loans to cash-strapped banks by half a percentage point, and this alleviated the pressure for a while.

A pattern seemed to be emerging: a series of lurches from crisis, response, respite, to new crisis. Bank of England Governor Sir Mervyn King warned in December of the dangers of a descending spiral. The crisis in the euro area was one of solvency not liquidity, he said, and the interconnectedness of major banks meant the banking systems and economies around the world are all affected. Only the governments directly involved could find a way out of this crisis, he observed.

In January 2012, Standard and Poor's downgraded seven nations including France.

Thereafter came a four-month respite in which equity markets rallied strongly and interest rates on bonds fell before the next lurch downwards.

Through all this, the financial institutions set on getting rich on carbon carried on plying their wares uninhibited by regulators in any way. In February 2012, a Canadian tar sands operator, backed by Chinese state-owned investors, raised half a billion dollars on the Hong Kong stock exchange. Cornerstone investors included China Investment Corp, China's sovereign wealth fund, and Sinopec Group, the country's biggest oil refiner. The operator had a name guaranteed to ruin my breakfast as I read about its success in the *FT*.

Sunshine Oilsands.

***

What do you do as the regulator if the financial system you oversee nearly drags the national economy into an apocalypse? You set up a committee, naturally. You charge it with making sure such a thing never happens again. The Bank of England set up a Financial Policy Committee in 2011 to 'contribute to the Bank's financial stability objective by identifying, monitoring, and taking action to remove or reduce systemic risks with a view to protecting and enhancing the resilience of the UK financial system'.

In January 2012, the growing constituency of people concerned about the carbon bubble wrote to the Governor of the Bank suggesting that the committee look hard at the risk of stranded assets in carbon. Twenty-one signatories spanned the financial services industry, non-government organisations, economics, policy and politics. They included the CEO of Aviva Investors, the chairman of Climate Change Capital, and the leaders of the main British environment groups.

Five of the top ten FTSE 100 companies are almost exclusively high carbon and alone account for 25% of the index's entire market capitalisation, we warned. This exposure is likely to be replicated in other indices, by companies, in bank loan books and in the strategic asset allocation decisions taken by institutional investors. At present, regulators are not monitoring the concentration of high carbon investments in the financial system and have no view on what level would be too high.

To understand the extent of the potential problem we need to assess global, and particularly UK and European, financial exposure to high carbon, extractive and environmentally unsustainable investments.

In February, we received a response from the Governor. The scenario we worry about, Sir Mervyn King said, would require three key ingredients. 'First that the exposures of financial institutions to carbon-intensive sectors are large relative to overall assets; second that the impact of policy and technology working to reduce returns in high carbon areas is not already being priced into the market, either through lower expected returns or higher risk premia discounting these returns; and third that any subsequent correction would take place over an insufficiently long period of time for the relevant financial institutions to adjust their portfolios in an orderly manner. The necessity of all three conditions being met raises a question in our minds as to whether or not this is a potential threat to financial stability.'

Disappointing. We believed all three conditions clearly were met.

But Sir Mervyn kept the door open. 'Nevertheless, there is clearly scope for further evaluation of these issues, in particular the potential scale of the risk and transmission mechanisms through which it might impact UK financial stability. To this end, we will endeavour to include this in the list of topics we regularly discuss with market participants, to assess whether or not this is a risk of which they are aware and the extent to which they are taking it into account in their investment decisions. In addition, Andy Haldane, the Bank's Executive Director for Financial Stability, would also be happy to meet with you and discuss the issues you raise.'

Bank of England, 20 April 2012

I have walked past the magnificent façade on Threadneedle Street many times, but never been inside before. Now I sit somewhere in the interior at a round table with Andy Haldane and a selection of those who signed the letter to the Governor.

A lot of good suits. A lot of British bonhomie. Terrible gender balance. A cross-section of the City of London, really.

Except the agendas are different. We explain our fears of a carbon bubble. We sketch policy- and technology-development scenarios that could begin to deflate the bubble. The longer we delay dealing with this problem, we observe, the larger the carbon bubble we have to wind down will get, the greater the chance of it bursting and causing a crisis, and of course the greater the chance of extreme climate events, themselves capable of extreme damage to the capital markets.

Haldane tells us that the Bank has already started to explore the issue with its constituency. It has added a carbon-asset-bubble question to its usual list of issues to discuss with its City contacts.

He asks whether there are more members of the financial community who are concerned about this, beyond the four in the room with us – Aviva, HSBC, PWC and Climate Change Capital – and if so what they are doing about it.

We respond that there is a sympathetic group who understand it but the current initiatives that exist do not tackle this kind of systemic risk. As individual organisations or even groups within the City it is impossible to tackle systemic risk, and the externalities keep getting passed up and down the investment chain without being factored in.

We use an analogy. Climate risks can be likened to pensions liabilities, which once didn't feature on company's balance sheets. Now they do. That's where we want climate risks.

Haldane seems to take that one on board.

He says he will prepare a note on our concerns and requests for the next meeting of the Financial Policy Committee of the Bank.

Kings Cross, London, 18 May 2012

The impressive theatre in the *Guardian*'s headquarters. John Elkington, corporate engager extraordinaire, has organised a one-day conference on what he calls 'Breakthrough Capitalism': the changes needed to re-engineer the system so that it has a chance of not crashing its way to ruin. He has asked me to give an energy entrepreneur's appraisal of modern capitalism based on my 12-year adventure in the markets.

I have been thinking hard about this one. It is time to be really frank, I have decided. If not now, after all that has transpired, then when? I worry that I have been pussyfooting around too much.

The content of this talk is not going to please some of my investors, or many of my peers.

I am nervous. I have given so many talks, over so many years. My pulse rate has long since ceased to rise. But today I feel a little like I did as a young academic, all those years ago, standing up in front of a roomful of professors. My voice shakes a little as I start.

Just relax, dammit.

To escape our increasingly costly oil addiction and the ruinous six-degree-hotter world we are heading for as things stand with our fossil-fuel dependency, we are going to need amazing companies to emerge: the cleantech equivalents of Microsoft, Google and their like. They must span the full family of renewables, in electric power, heat and motive power.

Problem is, as many venture capitalists are counting to their cost, where are they? Despite growing investment, something is holding back the cleantech equivalent of the digital and internet revolutions. For myself, after more than a decade on this frontier, I have come to believe it is something very dysfunctional.

Before I elaborate, and explore the implications, I've been asked to summarise my personal voyage through the markets, so that you can judge where I am coming from.

Solarcentury has been a lucky company relative to many of its peers. Lucky break number one: we raised our first capital in the same month the first crash of the twenty-first century began, the dot.com crash.

In the withering recession that followed, we struggled across the Valley of Death, surviving several near-death experiences along the way, making it to profitability in Year Six. Just. By then we had our first own products in production – solar electric roof tiles – and had completed a greater diversity of installations on buildings than probably any other solar company in the world. We were well on the way to the first of our four intra-company marriages. So if we do end up joining the mounting solar bankruptcies, hey, at least we've made it as a dating agency. And we set up our own charity, SolarAid, with 5% of our annual profits, as we had always intended.

In Year Seven, we took on more capital. Now we could invest to broaden our product range, strengthen our leadership team, go international.

Year Eight. The year of the credit crunch. Lucky break number two. We raised £13.5 million in August of 2007. The bankers stopped lending that same month. Without that cash we would probably be bankrupt by now, like so many of the solar companies we started out with.

It would have been the banks that would have bankrupted us, as they have bankrupted so many by dint of their systemic excesses, still as yet inadequately confronted.

Instead, we set off across the worst recession in a century on our second ascent to profitability. Today, in our Year 13, despite our industry's thin margins, we have more cash in the bank than we raised in our last round. We have kept improving and reducing the cost of our residential roof products, developed competitive products for all kinds of commercial roofs, and established a multiple-sourced supply chain in China controlled by an innovation team that has become a magnet for talent in design and value-engineering. We have installed big ground mounts: enough to boost our top line, not so many that we stress the cash flow.

What lies ahead for companies like ours? For those that can survive amid the current carnage, the rewards will be huge. A recent McKinsey study spells this out graphically. It estimates an economic potential of 1,000 gigawatts by 2020, eight years from now. This, as McKinsey say, will change the face of the global energy industry.

The driver is the spectacular plunge of solar costs at the same time conventional energy costs are soaring. Solar will be cheaper than conventional energy soon in multiple markets, and already is in some.

This will drive a new wave of financial innovation – green bonds, integrated energy-services models, and other things that mobilise the great prize of serious access to what is essentially our own money, in the pension funds.

So Solarcentury can live in hope, in the face of all this. We're not that elusive green Google our investors hoped for – but we are a success story compared to many peers, which is to say we are still alive. A key point, therefore, is that none of what follows is sour grapes. Well, not yet anyway.

So why are people like me gloomy about modern capitalism? Why do we think it needs to be re-engineered root and branch. Why are we in search of a renaissance in society?

Let's start with the small matter of modern capitalism having come very close to destroying itself of its own volition. And it may yet do so. We live daily with the ramifications of this little problem. Distressed clients in the construction industry: that's just the start.

Next, one of the many symptoms of modern capitalism's malaise. I call it the Big Energy Incumbency. I think of the Incumbency as a quasi-institutionalised human culture with a strong and indeed potentially deadly default self-defence mechanism, executed by three sets of actors.

Group one: many – but thankfully not all – who control the current energy system in the big energy companies. Nuclear and gas allow the energy giants to keep their customers away from the communal energy self-reliance that renewable energy can make easy.

Group two: many – but thankfully not all – who bankroll the current system. Conventional Capital tends to prefer Big Energy. Among other things, bonuses come easier.

Group three: many – but thankfully not all – of the institutional and ideological supporters of Big Energy in public service. Nuclear and gas allow civil servants and politicians to keep power centralised, literally and metaphorically.

OK, you might say, all this is regrettable, but why would you think the whole operating system is not just in crisis, as the *FT* concedes, but broken? Well, consider three more big worries. First, the system seems essentially incapable of responding to potentially existential threats.

Consider three risk issues I work on: climate change, oil depletion and – with my colleagues in the think-tank Carbon Tracker – carbon asset accountancy on stock exchanges.

Climate change is surely the ultimate litmus test for modern capitalism. If our operating system is so blind to genuine value that its default outcome is a six-degree-hotter world, what does that tell us about that system?

If our system allows us to go on entrenching oil dependency in economies knowing that oil is finite, and that production will peak, it's only a question of how soon; what does that tell us about that system?

If our system actively encourages us to go on clocking fossil fuels on our balance sheets and stock exchanges as assets at precisely zero risk of impairment, knowing that only around 20% of existing proven reserves need to be burned before we tip over into dangerous climate change, what does that tell us about that system?

These things tell us, I submit, that the system is dysfunctional to the point of being suicidal.

Big worry number two: the system is riven with short-termism. A short-termism that is frequently so grotesque that cartoonists lampoon it with vitriol in national newspapers.

Like most entrepreneurs, I experience this short-termism on a daily basis. Longer-term innovation is just one of the casualties. I worry that the pressure on investors for quick returns has been and is holding back cleantech innovation with breakthrough potential but longer gestation time frames. Yet the ultimate investors are often pension funds, controllers of the people's money, the very entities that should be most interested in the long term, in a sane system.

Big worry number three: the system allows the Incumbency to try to impress Goebbels in their defence of the status quo. The multi-million-pound, below-the-line PR budgets of many of the energy giants ensure that every occasional malfunctioning wind turbine will make national tabloid headlines. That people believe it is green measures that are inflating their energy bills rather than the wholesale price of gas. That every new oilfield find means that peak oil is scaremongering. That every unconventional gas exploration programme results in a new Qatar. That an injudicious e-mail by a climate scientist can be puffed up into such a faux Watergate that it can derail an entire climate summit. Self-defence this poisonous should be anathema to a civilised society.

So what happens next, in the face of all this? A new, re-engineered, capitalism is likely to emerge, I believe . . . because of one fundamental driver.

There will very likely be another major system shock soon. It might be intrinsic to the markets, a product of the vast multifaceted credit bubble that modern capitalism has built around the world. Or it might be the oil shock that the UK Industry Taskforce on Peak Oil and Energy Security and others warn of. This next shock is unlikely to prove suppressible by bailout, as the credit crunch was.

The people are not likely to allow a second shock to pass without forcing the political class to execute deep system reform.

Localism and the democratisation of energy, already growing fast ahead of the crash, will be driven even harder by expediency.

Society will finally be forced to pick the energy path that the Incumbency strives so hard to keep us off. The survival technologies and tactics will have to be mobilised fast, via green new deals in multiple countries, financed at last by pension funds acting as though people should have a viable society into which to retire.

I believe that if we can do this, a huge prize then materialises: a renaissance in society.

Can we do this?

Yes.

Will we do it?

I don't know. But we are going to find out, and we have to try.[4]

Chapter 16

A new era of fossil fuels

I saw a remarkable thing two decades ago: 152 world leaders in one hall, milling around waiting to be photographed together on a tiered stage like pupils assembling for a school photo. They were in Rio de Janeiro for the 1992 Earth Summit. They went because public concern about the deteriorating state of the global environment, and the continuing failure to tackle global poverty, was high around the world. They left with a clutch of conventions aiming to fix these problems. Those treaties were long on good intentions, short on teeth in the form of legally binding commitments. But there was a strong scent of hope in the air that day, in that hall.

In June 2012, the Rio-Plus-20 Earth Summit was supposed to beef up the agreements signed two decades before, and others besides. But the key world leaders from the developed world didn't even bother to turn up. Instead they left matters to their ministers.

There were two main reasons for this dearth of global leadership. On the one hand, heads of state had been badly burned by the embarrassment of Copenhagen. On the other, the financial crash had left fewer citizens pressuring them about the environment and development. In the developed world, many people were now worrying more about whether they could pay energy bills, meet mortgage repayments, and such issues.

All the summiteers of 2012 could produce was a plan to set sustainable development goals at some unspecified date in the future on themes yet to be decided. An open working group of 30 nations was sent away to report back in September 2013 on what themes should be discussed.

Horrified environment and development organisations pondered how best to project the full magnitude of this lost opportunity. They tended to opt for the obvious theme. Global leaders had set a new definition of hypocrisy. They had just bailed out banks to the tune of trillions of dollars. They were funnelling a trillion dollars in subsidies each year direct to fossil fuels, vastly more than they directed at renewables. They seemed to have decided that efforts to stop environmental crisis and eradicate poverty were worth only the merest fraction of those trillions, and little or none of their collective and individual political capital.

***

The oil and gas industry's campaigning, meanwhile, was achieving some remarkable results. In May 2012, after an intense industry lobbying campaign, an €80 billion EU programme rebranded gas as green energy. This created a totemic precedent for oil industry PR. It also meant gas could attract funds that might otherwise have gone to renewables.

The same week, the International Energy Agency produced a special *World Energy Outlook* report on unconventional gas. Its central premise was that shale gas could lead the way to what the agency called 'a golden age of gas'. There was one proviso, the agency said. The industry would have to make sure it worked hard to address public concerns about the social and environmental impacts of drilling.

Those questioning this exuberant narrative now included one of the biggest investors in gas, Scottish Widows Investment Partnership, which released an analysis suggesting gas is no better than coal, in terms of greenhouse-gas emissions, when methane leaks from gas-industry operations are taken into account. So-called 'green completion' technology is available to companies, but most companies fail to use it, especially in the USA, because it costs them to do so and because there are no penalties for leaking gas into the atmosphere. Shockingly, shale gas and its fugitive emissions now account for fully 20% of all US greenhouse-gas emissions.

Concerns about the amount of debt being raised to finance shale gas drilling continued to grow. In June, an *FT* columnist echoed some of the warnings that had appeared in the *Wall Street Journal* the previous year. The state of play reminded John Dizard of the sub-prime mortgages fiasco. Shale-gas producers were spending up to five times their operating cash flow to fund their land, drilling and completion programmes, all in the face of enduringly low gas prices. 'Wall Street should have provided reality checks to the shale gas people', he wrote. 'Instead, they just provided cashier's cheques with lots of zeroes at the end.'[1] It was worse than a real estate bubble, wherein construction could just slow or stop while the courts unscrambled the mess. But America had become dependent on gas-fired power. How could the process stop? Drilling would have to continue and, given the steep decline rates of shale gas wells compared to conventional wells, would in fact have to accelerate if current production levels were to be maintained. If the cost of this future drilling and the servicing of existing debt were to be met, gas prices would have to adjust upwards. Substantially.

Art Berman, a long-term gas-industry insider critic saw the problem even more starkly. 'Once you start drilling shale wells you can never stop', I heard him tell the annual meeting of the Association for the Study of Peak in Vienna. 'Shale plays are not a renaissance or a revolution. They are a retirement party.' We were watching an industry, in his view, 'self-destructing'.[2]

The tale of *Alice in the Looking Glass* looked to be increasingly relevant. And in the detailed researches published on the *Oil Drum* website and elsewhere, analysts looking at the decline rates in shale gas and tight oil wells, and the implications for maintaining production, began referring in 2012 to a 'Red Queen's Race'. In *Alice in the Looking Glass*, the Red Queen is a character who professes: 'It takes all the running you can do, to keep in the same place.'[3]

In June, Exxon Chief Executive Rex Tillerson broke from the previous company line that it wasn't being hurt by low natural gas prices. 'We are all losing our shirts today', he told the Council on Foreign Relations in New York. 'We're making no money. It's all in the red.'[4]

But there was still money enough for the oil and gas industry's largest-ever campaign to sway a presidential election. In the run up to the November 2012 election the industry threw its weight squarely behind the Republicans, and a vision of an American energy landscape transformed by domestic gas and oil. The American Petroleum Institute's 'Vote 4 Energy' campaign aimed to increase access and limit regulation for oil and gas companies across the board, and especially in key states like Ohio, where fracking was soon to begin.

***

The financial crisis had for a while now increasingly resembled a rotten onion, the layers of which were peeling away one at a time. In June 2012 came a particularly malodorous unveiling. Barclays admitted it had long been rigging the London Interbank Offered Rate (Libor), the average rate at which banks supposedly lend to each other. Many other banks had been doing the same, it soon transpired. The scam was systemic.

Much commercial activity in the financial sector depends on the Libor rate. It is calculated daily from the rates banks submit, and once posted is used as the basis for setting the rates of loans, credit cards, and so on. Trillions of dollars in derivatives contracts are based on the rate.

The breathtaking truth was that banks had been colluding to misrepresent the rate at which they lend to each other, both to make money in their various casino activities and to misrepresent the true picture of their parlous financial state.

Big as the backlash had been since the financial crash, lava of a whole new indignant viscosity now erupted. 'This was market-rigging on a grand scale', an editorial in the *Financial Times* raged. 'It is hard to think of anything more damning – or more corrosive of the reputation of capitalism.' The *FT* considered it 'absurd that in the UK almost no bankers have been prosecuted for their role in the crisis and its fallout'. Now, with this new kind of systemic confidence trickery, no stone should be left unturned. As for Diamond, 'he was clearly responsible for its hard-driving culture. If he had an ounce of shame, he would immediately step down.'

But the broom had to sweep much further, in the opinion of the vituperative leader writers. 'It may therefore be necessary to retire this generation of flawed leaders.'[5]

I had never thought to read this kind of language in the pink paper in my lifetime. If this is what *their* editorial writers were thinking, I wondered, what must it be like over at the *Socialist Worker?*

Certainly, the Financial Services Authority and Serious Fraud Office – understaffed for so long in the era of light-touch regulation – would now be cancelling holiday far into the future.

Lower Marsh, London, 2 July 2012

It is one of those days in and around the office. The meetings stretch ahead back to back and someone seems to have forgotten the travel time between them. Next up, the lead author of the McKinsey 'solar-will-change-everything' report is in to give our senior team his views between the lines. The trade-off is that we have to tell him what it looks like under the bonnet of a $100 million downstream solar company in the insanity of the current markets. After that I head to Parliament, to address a committee of MPs holding a workshop on green energy jobs. Not everyone over in Westminster has been seduced by the lure of endless cheap fracked gas, it seems. I will have to take time out of that meeting to do my first ever joint pitch on behalf of Solarcentury and SolarAid. I will be speaking down a mobile phone from a corridor in the House of Lords to senior executives of a giant corporation, in multiple countries, trying to persuade them both to stack their vast acreage of roofs with solar and, at the same time, to help us accelerate channels to mass markets for solar lighting.

Now I am on the second floor of Solarcentury's office, listening to a presentation on China. The entire company is here, casually dressed people sitting in chairs or leaning on desks, trying to ignore the banter and cooking smells wafting in from the market street outside. We are listening to the amazing Peggy Liu, founder of the Joint US–China Collaboration on Clean Energy. Peggy, a Chinese-American ex-venture capitalist now resident in Shanghai, is telling us her hopes for collaboration between the two biggest greenhouse-gas emitter nations in accelerating the technologies and strategies that can fix the climate problem. She is off the record, under Chatham House Rules, and talking deep realpolitik.

The team listens avidly. Some of them spend long periods in China tending our complex supply chain. All of them depend for their livelihoods on that supply chain. And Peggy is a force of nature, an utterly compelling speaker. She has them rapt.

She is treading in illustrious footsteps. Her seminar is part of a series in which I invite eminent external folk to talk off the record to the team. Anita Roddick stood where Peggy does now, shortly before her untimely death, telling the back-room story of her creation, the Body Shop. Charles Dunstone, founder CEO of the Carphone Warehouse, recounted hilarious and unrepeatable stories of his early days. Ron Oxburgh, a former Shell chairman, told of his brave efforts to reform an energy giant from within.

Peggy is talking now about corruption: ours and theirs. Smiles are appearing on faces. Heads are shaking.

I am only half-listening. Today I am wrestling with a new level of incredulity about the corruption a few miles away in the City of London. Bob Diamond and his den of knaves have been caught with their collective hands jammed in the till. Yet still he has not resigned. His chairman has, but Diamond is trying to tough it out.

I have also learned that Bob Diamond's right-hand man is called Rich Ricci. Did anyone ever need more proof that God has a sense of humour?

I look around me at the team that is being battered so hard by the shakeout from the excesses of both the financial and the energy incumbencies. They work long, hard hours, with no hope of the kinds of salaries, or bonuses, that the hit men of the financial and energy incumbencies trouser.

Diamond has talked of the time for remorse being over. He has spoken of the importance of culture in a bank. 'For me the evidence of culture is how people behave when no one is watching', he said in a recent speech.[6]

Priceless.

What cynicism. What a system.

I refocus on Peggy's words. She is pushing the envelope in trusting her listeners.

It doesn't sound as though the Chinese are doing too well with their system either.

Bob Diamond was fired the next day, but the use of extreme language in normally conservative institutions continued. The deputy governor of the Bank of England, Paul Tucker, suggested that banking had become a 'cesspit' out of which another Libor scandal could easily emerge.

HSBC had avoided much of the muck that the other big banks had attracted up to now, but on 17 July it took its turn in the cesspit's rogues' gallery, forced to admit that it had been laundering money for Mexican drug cartels, terrorists, pariah states and organised crime. Charles Ferguson, director of a film on the financial crash, *Inside Job,* summarised the state of play starkly: 'Banking has become criminalised in a way that threatens global stability.'[7]

In the face of this wall of wrongdoing, the grandees charged with chairing the various committees that had been set up to recommend reforms to the financial sector pressed ahead with their work. On 23 July, the Kay Review of equity markets made multiple good recommendations for improvement. Pundits anticipated immediately that only a few could make it to legislation, however. In September, the Vickers Review of banking recommended giving the banks until fully 2019 to ring-fence their high-street operations from the casino arms.

***

By July 2012, problems with the oil and gas industry's shale operations were proliferating in terms of both the environment and balance sheets. The Proceedings of the National Academy of Sciences published a study suggesting that toxic fracking fluids were seeping upwards thousands of feet into Pennsylvania's drinking water supplies. This was the kind of thing likely both to galvanise local opposition and to build pressure for more and better monitoring. It also tended to undermine the support chorus engendered by industry funding. Industry-funded academics were at this time facing ethics investigations over pro-fracking studies at universities in Ohio and Texas.

An article appeared in the *Wall Street Journal* in which an independent research firm questioned the size of gas giant Chesapeake's reported shale-gas reserves. And within a few weeks of that setback, respected industry research firm Bernstein suggested that the much-hyped Bakken formation of Montana was heading for an oil production crash because so much of the easy terrain had been drilled. Bernstein analyst Bob Brackett wondered if the entire Bakken formation, including North Dakota, could be in trouble too. 'There is an emerging view of a wave of oil production (from shale and otherwise) coming', he wrote. 'I just want to point out the difficulties in an exuberant view.'[8] The *Oil Drum* website was crammed with analyses such as this by now.

It was striking, however, to see how much more the industry's exuberant narrative echoed in the media than any questioning of it did. One particularly successful industry gambit involved a former oil executive now at Harvard University. Leonardo Maugeri posited that the next decade could see the largest increase in oil production capacity since the 1980s, with global production reaching more than 110 million barrels a day by 2020, leading to prolonged overproduction and a long-term downturn in oil prices. This was the beguiling cornucopian message writ large. BP had sponsored his work.

Counter-views quickly appeared in the obvious places, such as the *Oil Drum* and other specialist websites. But they achieved only a fraction of the mainstream media air play that Maugeri's message did.

David Strahan, an ex-BBC journalist active in the Association for the Study of Peak Oil, interviewed Maugeri by phone and e-mail, and forced him to admit to a basic mathematical howler in his core argument: incredibly, an incorrectly compounded percentage. Strahan published this revelation in *New Scientist* magazine. It did little to dent the Maugeri mantras echoing in the mainstream press all around the world.

The oil industry's PR effort held the ability to scare those of us watching closely. Industry executives and PR advisers gathering to strategise in Houston spoke of a 'war on shale gas' by an anti-fracking 'insurgency'. This kind of language suggested an entrenched group-think incapable of distinguishing, on the one hand, between concerned citizens and analysts in a democracy and, on the other, al-Qaeda and the Taliban.

Others in the industry were more civilised, but only a little less scary. In August it emerged that Shell had run a two-day course on their view of the future for dozens of senior and mid-level civil servants across ten Whitehall departments and agencies. Such access to officialdom was something the renewables industries could only imagine. What view of the future would be offered on this course? Would a high-renewables scenario have featured? If so, how would Shell be explaining the fact that it had essentially pulled out of all renewables except biofuels?

The US shale gas boom was by now hurting the conventional gas producers. In August 2012, President Putin announced that Russia was pulling out of the giant Shtokman gas project in the Arctic. This, one of the biggest gas-fields ever found, had been rendered uneconomic for the foreseeable future, the Russians said, because of soaring costs, falling European demand and cheap shale gas in America.

The US shale gas boom and the exuberant narrative being weaved around it were by now making converts in high places outside America. In the UK, Chancellor George Osborne joined them. In July, he wrote a letter to the Secretary of State for Energy and Climate Change, Ed Davey, proposing a deal whereby the Treasury would support a workable reduction in the subsidy rate for onshore wind if Davey would essentially put the brakes on wider renewables in the UK. The letter was immediately leaked, no doubt by an outraged official in Davey's office. What the Chancellor feared, his missive revealed, was that too much progress in the renewables markets would put off the investors needed to turn the UK into what Her Majesty's Treasury wanted to see emerge: a 'gas hub'.

If the renewables industries in the UK hadn't known just how badly the cards were stacked against them by this time, they surely did now.

***

Premature peak oil remained a non-topic in the UK, as in so many other countries. This was true even when the warnings came from right within the British government's own political tent. In August, the international business editor of the *Daily Telegraph*, the main pro-Conservative newspaper, reviewed the state of play in oil markets. If no energy crisis was in prospect, Ambrose Evans-Pritchard wrote, then why are oil prices so persistently high? His conclusion: 'Peak cheap oil is an incontrovertible fact.'

In September, Citi produced a report on Saudi Arabia's domestic oil-consumption problem. Oil and its derivatives were being used for about half of the kingdom's electricity production, the bank reported, which at peak rates was growing at about 8% a year. If this continued, the kingdom risked becoming an oil importer within just 20 years.

Evans-Pritchard phoned me for a comment and printed it the next day. Britain is sleepwalking into a potential disaster by failing to prepare fully for a global supply crunch, I said. Their refusal to listen to warning signals is comparable to the complacency in the build-up to the financial crisis, but with graver implications for the British economy.

The *Telegraph* reporter hazarded his own response to this: 'I agree.'

This put the UK Industry Taskforce message right under the noses of the Conservative leadership, in a paper that was in many ways a newsletter for the party.

***

The Treasury was too busy plotting gas nirvana with the oil industry to notice, it seemed. And the Department of Energy and Climate Change was too busy protecting what remained of the nuclear industry. In July 2012, it emerged that they had colluded with Eon and RWE to soften the impact of the latter's withdrawal from nuclear.

The media-management challenge for the nuclear lobby was becoming severe indeed. On 5 July a Japanese parliamentary panel came to the conclusion that the Fukushima disaster was man-made. Industry advocates had long maintained that the accident could only have happened because of the devastating tsunami that hit the plant. But the parliamentarians now said they could not rule out the possibility that the earthquake preceding the tsunami had triggered the chain of events leading to the reactor meltdowns before the wave struck. This had been made possible, they said, by poor regulation and collusion between the government, the operator and the industry's watchdog.

Whatever the Fukushima forensics, the nuclear reactor fleet worldwide was a net 12 reactors down in 2011. Only seven new reactors had started up, and 17 had been permanently shut down. Of Japan's 54 reactors, many more might now follow. By July, only one had been turned back on.

The industry badly needed some good news to reverse the constant stream of negativity around the world, but on 16 July the Olkiluoto reactor in Finland was delayed yet again. The facility had been due to come online in 2009. Now it would not be turned on before 2014.

Organisations hitherto supportive of nuclear power continued to distance themselves from the potentially sinking ship. Said Jeff Immelt, GE CEO, to the *Financial Times* in an interview: 'When I talk to the guys who run the oil companies they say, look, they're finding more gas all the time. It's just hard to justify nuclear, really hard. Gas is so cheap and at some point, really, economics rule. So I think some combination of gas, and either wind or solar . . . that's where we see most countries around the world going.'[9]

Japan, the world's third largest user of nuclear power before Fukushima, had planned to increase nuclear massively by 2030. On 14 September 2012, it announced it would be phasing out all its reactors by 2040.[10]

***

On 24 July 2012, scientists reported that the surface of the Greenland ice cap was melting so fast that when they first saw the satellite measurements, they thought the instruments must be faulty. But no, that month the ice was melting at a faster rate than at any other time in recorded history, with virtually the entire ice sheet showing signs of thaw.

In August, data from the first purpose-built satellite launched to study the thickness of the polar caps was published. It showed that Arctic ice was disappearing at a far greater rate than expected.

In September, the UK's top ice expert, Professor Peter Wadhams of the Scott Polar Institute, predicted that final collapse of the Arctic sea ice was just four years off.

FT Global Energy Leaders Summit, London, 18 September 2012

Exxon, US Coal, BP – or rather ex-BP in the shape of Tony Hayward – and me. A panel at an *FT* conference, senior correspondent Ed Crooks in the chair. The *FT* has chosen a title for the session to fit the times: Are We Entering a New Era of Fossil Fuels?[11]

They could never have called it 'Will Cleantech Seize the Energy Incumbency's Markets?' Their business model wouldn't have worked. Nobody from the fossil-fuel sector would have showed up. They would have been left with a few hard-up renewables types and the odd frustrated investor.

The lady who runs Exxon in Europe, Linda du Charme, opens. Yes, we are in a new era of fossil fuels, she says. The world will still be 80% fossil fuel-powered in 2040, because they're here, they're versatile, they're affordable. Unconventional gas and unconventional oil have been a game-changer in America, and they will in other countries too.

I wonder at the mindset that allows an intellectual analysis like that.

They are here, they are versatile, they are affordable.

That's it? That is all we are capable of? We, the species that can do Mozart? With the heat-trapping ability of greenhouse gases in our thin atmosphere undisputed when all this stuff is burned? With the shock and awe of the melting ice cap plain to see in the newspapers this morning?

John Eaves, CEO of giant American coal company Arch Coal, also thinks we're entering a new era of fossil fuels. He doesn't want coal to be left out. Coal has been under attack in the US, he says, but there is plenty of growth potential.

Tony Hayward next. He looks a bit red in the face. So do I. His colour is from sun and sailing, I imagine. I can feel by this point that mine is soaring blood pressure.

Hayward waxes lyrical about how much oil there is in Iraq, if the political problems standing in the way of his industry can only be overcome.

These days he is CEO of an oil company targeting Iraqi oil. Investors have been queuing to back him.

We have never left the fossil-fuel era, Hayward continues. Absent intervention to do with other things, like climate change, which I'm sure Jeremy will want to talk about, it remains the lowest-cost form of energy.

Ah, finally a mention of climate change. Fleeting, dismissive, but a mention.

And the choice of word. Intervention. Not policymaking, or even essential policymaking. Intervention.

It's clear today that there is infinite gas, Hayward says. There's probably infinite oil, certainly infinite in the timescale that many of us can think about. There's certainly infinite coal.

Now I am being transported to Stranger-Than-Fiction Land again. Tony Hayward, like me, is a PhD geoscientist. He knows that it is physically impossible for oil and gas to be infinite. Hydrocarbons need organic matter to form, over millions of years. Only so many sedimentary basins had the right conditions. Years earlier I had heard a BP geologist tell a conference that his company 'knew where the source rocks are'. There are only so many of them, he had said.

I marvel at the default myth making. That's what you have to be capable of, I think to myself, to do what he does. You have to be able to spin myths with a straight face and smile to your peer group. And to do that, you probably have to be capable of swallowing myths yourself.

We will find a better way of producing energy at some point, Hayward is now saying, but it's not any time in the next few decades.

I'm going to have to bring you in now Jeremy, says Ed Crooks, laughing.

We are thirty minutes into the session. If I had a blood pressure monitor clamped to my arm, I would surely be breaking world records.

Well, I say. Absent intervention. That's what this is all about.

Chapter 17

More unhinged by the week

In November 2012, Chesapeake wrote its gas reserves down. Its debts stood at more than $16 billion. It was now only a matter of time before once high profile CEO and founder Aubrey McLendon would be ousted by his investors. This duly transpired in January 2013.

The US gas boom had meant low prices for consumers and good profits for the bankers handing out all the loans to the frackers. But for the drillers it was a very different story. In the previous four years, the top 50 oil and gas companies had raised and blown an annual average of $126 billion on drilling, land acquisition and other capital costs. Could it be that the boom was entering a time of consequences? Even the American Petroleum Institute seemed to have its doubts, speaking only of 'an opportunity' for the boom to continue. The API saw risk of tight regulation in new rules due in 2013. Others saw risk that unignorable health impacts might emerge, that water shortages could constrain operations, and that long-term production might fall below expectations. The level of civil protest in America was rising.

Even if domestic risks could be faced down, there were substantial risks that the shale gas story would not be exportable beyond the continental United States. On 2 November, the *Wall Street Journal* ran a rare article summarising the views of those doubting the incumbency. 'Global gas push stalls', it read. 'Oil companies are running into obstacles as they try to replicate the U.S. experience on other continents. The result is that significant overseas shale energy production could be a decade away.'

A decade away. This was beginning to sound like the 'nuclear renaissance'.

There are five main reasons for this view. First, mineral rights are owned in a fairly unique way in the USA. The land is mostly owned by private landowners, most of them easily persuadable to go with the flow for the riches they can earn. In most other countries, the mineral rights tend to be owned by governments. Second, the environment concerns will play much more strongly in many countries than they do in the USA. Third, the geology in other nations often compares unfavourably with the US shale basins, as Exxon had found in Poland. Fourth, the dire water-intensity of the fracking process is becoming increasingly problematic in a water-constrained world. This seems to be the main reason why fracking is progressing so slowly in China, for example. Fifth, many countries lack the infrastructure to drill and transport gas and oil. The US, in contrast, has been producing large amounts of oil and gas domestically, in free markets, for more than a century. A lot of pipe has been laid, as it were.

The UK Chancellor announced in December 2012 a gas strategy for Britain that would entail up to 37 gigawatts of new gas plants, provided national carbon targets set for the mid-2020s could be reined back. He had earlier announced that he intended to give generous subsidies to gas companies so they could frack their way to cheap gas under Britain. 'Chancellor backs gas to fire Britain up', read the *FT* headline on 3 December. It sat very uncomfortably with the *Wall Street Journal*'s headline only the day before.

For a country like the UK, with a severe austerity programme under way, the stakes could not be higher. There will probably be only one shot at capitalising a twenty-first century energy infrastructure. If multiple tens of billions are poured into gas infrastructure, to the detriment of renewable and energy-efficiency alternatives, and then the shale gas vision proves unexportable from the US, national disaster awaits. Lights go off. Fuel poverty soars. Any semblance of national economic recovery stalls. If this pattern is replicated simultaneously by multiple governments unwisely buying the mythology built around the US shale gas boom, then disaster could easily assume the proportions of a global crash.

***

Should it indeed take a decade for the US shale gas phenomenon to prove exportable to any significant degree – provided, that is, that the bubble does not burst in the interim in the US itself – what chance is there that nuclear could speed up the timing of its next generation reactors? In October 2012, Areva and its Chinese partner pulled out of their bid for RWE and Eon's nuclear assets in the UK. EDF, standing alone now with Centrica, threatened that it would not build UK nuclear plants without major subsidy. The same month, to the UK government's great relief, Hitachi bought RWE and Eon's nuclear assets, starting what they called a 'hundred year commitment' to nuclear in the UK. But then, on 3 December, EDF was forced to announce that the cost of its flagship reactor at Flamanville had risen yet again, to fully €8.5 billion, and that it was further delayed and couldn't come onstream before 2016. The price ticket had gone up by €5.2 billion on top of an original €3.3 billion, and the build time by four years on top of an original five years.

Italian energy company Enel immediately pulled out of its share in the Flamanville project.

'The last 24 hours have killed French nuclear', a UBS analyst wrote.

If it was dead in France, the lead nuclear nation, a state with a record 80% of electricity coming from nuclear, what chance did it have in the UK and beyond now?

Nuclear advocates tended to the view that China would be pressing ahead anyway, but actually the Chinese had only just restarted their nuclear programme after a year-and-a-half post-Fukushima shutdown and safety review, and professed only to be building 'a few' reactors between 2012 and 2015.

Centrica was next to pull out of nuclear power in the UK, in January 2013, writing off £200 million in its partnership with EDF in the process.

In February, the now desperate EDF CEO Henri Proglio made the amazing assertion that unless the UK government guaranteed him profit from nuclear operations, he would walk away. 'We won't do it', he said, if the price for the power isn't high enough. 'I have no reason to take the pressure off the people I'm talking to.'

I sent a spoof tweet on hearing this latest outrage, threatening that unless Her Majesty's government guaranteed Solarcentury solar profits, we would quit the country.

And four days later came news that Areva's Finnish reactor, the sister of the reactor they were building for EDF at Flamanville, would be delayed yet again, to seven years beyond the original schedule.

When I read on 18 February that the UK government was indeed considering giving EDF guaranteed profits for 40 years if they would but build a few nuclear power plants in Britain, I felt there was a certain inevitability about it.

***

In September 2012, Total announced that it would not be drilling for oil in the Arctic, because the risks of a spill were too high. UK MPs at the time were demanding a moratorium on all Arctic drilling, and proposing unlimited liability for any companies who did drill there. They had heard what they considered to be compelling evidence that if a blowout occurred just before the dark Arctic winter, it would not be possible to cap any oil spill until the following summer. If other companies were to do the same as Total, this would narrow the field for projected significant new global oil production essentially to unconventional oil and deep-water non-Arctic oil.

Much depended on Shell now if the oil industry was indeed to open up the supposed oil riches below Arctic waters. They were given clearance for limited drilling in August 2012 and immediately ran into trouble. In September, they were found to have submitted crucial spill containment equipment to only scant testing. By January, both their drilling rigs were out of action with technical problems. In February, they put their Arctic operations on hold for a year.

What might this mean for peak oil? The IEA had for some years now been rocking the incumbency boat somewhat with its warnings about dwindling oil production. But the 2012 *World Energy Outlook*, produced in November, reverted to historical type, as already mentioned. In it, the IEA projected an oil price in 2035 not much higher than today's, with production at nearly 100 million barrels a day. The shale gas boom, the agency suggested, was spilling over into tight oil production to such an extent that the USA could reasonably expect to become almost self-sufficient in oil and gas by 2035. The USA would be the new 'Saudi America'. This was a quite remarkable turnaround.

BP beat the drum alongside the IEA. In January, its latest energy outlook showed all net growth in global oil supply to 2020 as coming from unconventional oil. The shale revolution would spread to other parts of the world. Global output of shale gas would treble by 2030. Tight oil production would grow sixfold by then. Asia had an estimated 57 trillion cubic metres of technically recoverable shale gas resources and 50 billion barrels of tight oil, compared with North American equivalents of 47 trillion cubic metres and 70 billion barrels. Bob Dudley told the press that peak-oil 'theories' were 'increasingly groundless'. David Frum translated this into tabloid speak on CNN.com: 'Peak oil doomsayers proved wrong'.

It was increasingly incredible to me how easily the media swallowed and echoed this messaging from BP. The oil giant was about to go to court with the US Department of Justice over the Deepwater Horizon spill. Had people forgotten so quickly how unreliable the corporation's word had been during that disaster? In February, the trial began, and the prosecution's evidence came spilling out. One internal BP e-mail, from the man in charge of the Deepwater Horizon rig for the company, had warned his superior onshore, just days before the fatal blowout that killed 11 men and made Tony Hayward wish he could have his life back, that staff were operating in 'chaos, paranoia and insanity'.

This seemed entirely allegorical to me as I read it. I wondered how many others were having the same feeling.

As usual, many dissidents responded with critiques of the IEA and BP view. Mostly these appeared on specialist websites and didn't reach the mainstream media. At the annual American Geophysical Union meeting in December, many American earth scientists warned that the idea of Saudi America was a complete myth. Raymond Pierrehumbert, Professor of Geophysical Sciences at the University of Chicago, wrote a summary of the American Geophysical Union's dissection of the faulty threads in the cornucopian narrative.[1]

Only a tiny fraction of the resource that the IEA emphasised was recoverable. In the Bakken shale of Montana and North Dakota and the Eagle Ford shale of Texas, which accounted for most of the current surge in US oil production, only 1 to 2% was ultimately extractable. This would amount to perhaps a two-year oil supply for the United States at 2011 consumption rates. 'That's significant', wrote Pierrehumbert, 'but not a game-changer. Even if it were to prove possible to achieve production rates comparable to those of Saudi Arabia, that would only mean that we would deplete the resource faster and bring on an oil crash sooner.'

And what would it take to ramp up production to such high but non-game-changing levels? Data presented at the AGU meeting showed America was now drilling 25,000 wells per year just to bring production back to the levels of the year 2000, when only 5,000 wells per year were being drilled. The new wells were expensive, costing some $10 million each in the Bakken, and their production rates dropped rapidly.

Pierrehumbert used the Red Queen's Race epithet: 'you have to keep drilling and drilling and drilling just to keep your production in the same place. At several million dollars a pop, that adds up to a big annual investment, and eventually you run out of places to put new wells.'

This analysis was much the same as the one Art Berman and others applied to the ultimate fate of US shale gas. Shale gas too was a Red Queen's Race, Pierrehumbert concluded.

In January 2013, Euan Mearns and Rembrandt Koppelaar summarised very instructive data on drill rig disposition in the USA, compared to the rest of the world. In January 1995, there had been a total of 1,738 oil and gas rigs drilling, excluding the former Soviet Union and China. By February 2012, there were 3,850, more than twice as many. Global crude plus condensate plus natural gas liquids grew from 68 to 84 million barrels per day, an increase of 16 million barrels a day of capacity, as a consequence. In January 1995, 42% of the world total of oil and gas rigs, excluding the FSU and China, were drilling in the USA. By October 2011, the figure was 55%. The tight oil drilling spree had produced less than two million barrels per day of new US oil capacity after deployment of more than half the total non-FSU non-China global drilling effort. Does this sound like a scenario that can lead the Yergin and Maugeri charge to way over 100 million barrels?

Oil drilling has recently turned down in the USA, Mearns and Koppelaar observe. 'If shale oil production was to continue rising into the future, we would expect to see the rig number continuing to go up', they note. 'It remains to be seen if the recent downturn in US oil drilling is temporary and what reasons lie behind this reversal.'[2]

In February, two damning reports on the shale boom, both from highly authoritative sources, hit the specialist websites, while missing most of the mainstream press. David Hughes, a geologist who had worked in the oil industry for four decades, three of them with the Geological Survey of Canada, wrote the most detailed survey of shale gas and tight oil production yet, using the government's own figures for 65,000 shale gas wells from 31 production areas. He showed that production had been on a plateau since December 2011, as the Red Queen Race took shape, and the best shale drilling areas, like the Haynesville, went into decline. The high decline rates of wells had meant $42 billion of capital had been needed to drill 7,000 wells, against a gas sales value of $32.5 billion in 2012, he observed. What kind of business was this?

Where was the basis for the hype, Hughes asked, when US gas production was growing overall by only a tiny amount? The decline in conventional gas production was barely being compensated for by the explosive rise in shale gas production to 40% of total gas produced.

As for tight oil, 80% of the rapidly rising production came from two drilling areas, the Bakken in North Dakota and the Eagle Ford in southern Texas, where more than 1,500 wells are needed annually to offset declines. Hughes expects these two areas to collapse back to 2012 levels of production by 2019, meaning that the rich heart of 80% of the US tight oil production would have amounted to a bubble of only about ten years' duration.[3]

In the second report, an ex-investment banker, Deborah Rogers, came up with a theory to explain the hype. She had worked for several Wall Street banks, including Merrill Lynch, had served on an advisory committee within the Department of Interior, and also on the Advisory Council for the Federal Reserve Bank of Dallas. Her background had led her to suspect that elements of Wall Street were playing games again, as they had in the dot.com era, and in the run up to the credit crunch.

Mergers and acquisitions (M&A) among shale companies had amounted to $46.5 billion in deals in 2011, making them one of the largest bonus spinners for some Wall Street investment banks. Yet shale wells were underperforming loss makers, in dollar terms at the time. So why had analysts and investment bankers been some of the most vocal proponents of shale exploitation, Rogers asked? They had helped to ensure that production continued at a frenzied pace: that the Red Queen Race kept running. In doing this they had created a glut in the market for natural gas. With supply of natural gas exceeding demand by a factor of four in 2011, gas prices had been driven to new lows. This had forced cash-flow-constrained companies to merge, so opening the door for M&A deals worth many billions to the bankers involved. They had also been generating huge bonuses from the debt requirements of the players.

Knowing her former colleagues as she did, Rogers felt compelled to conclude that they were rigging markets again. As for the oil and gas companies active in shale, they were more than willing players: they had overestimated their shale gas and tight oil reserves by a minimum of 100% and by as much as 400–500%, according to actual well production data filed with the states, Rogers alleged.

She also pointed out similarities with the market distortions in the run up to the credit crunch. 'As prices plunged,' she explained, 'Wall Street began executing deals to spin assets of troubled shale companies off to larger players in the industry. Such deals deteriorated only months later, resulting in massive write-downs in shale assets. In addition, the banks were instrumental in crafting convoluted financial products such as VPPs (volumetric production payments); and despite the obvious lack of sophisticated knowledge by many of these investors about the intricacies and risks of shale production, these products were subsequently sold to investors such as pension funds. Further, leases were bundled and flipped on unproved shale fields in much the same way as mortgage-backed securities had been bundled and sold on questionable underlying mortgage assets prior to the economic downturn of 2007.'[4]

This latest development shone a whole new light on the shale debate, and the scope for a market 'surprise'. How did an analysis like this play in the UK Treasury, where ministers were desperate for the shale fairy tale to be both real and exportable? Did news of it even reach their desks?

As for the International Energy Agency, they were no longer even in the business of coded warnings, it seemed. On 7 March, the new IEA Director-General, Dutch diplomat Maria van der Hoeven, wrote an op-ed in the *Financial Times* that made me laugh out loud as I read it. She came across like a full-on lobbyist for the oil and gas companies.

The US must avoid the shale boom turning to bust, she said. How to do that? Make sure all the transportation bottlenecks are removed so that North American crude doesn't have to trade at a discount to the rest of global oil, as is the case at the moment. The US needed more pipelines to get the oil to coastal refineries so it can be exported and traded at the same high price OPEC gets for their oil.

Clearly, she was worrying about the balance sheets of the US oil and gas companies, and their ability – like Chesapeake – to service their enormous and growing debts, and somehow break out of the Red Queen's Race.

Is it just me, I asked on Twitter, or is the energy world getting a little more unhinged every week?

***

The financial world had long since exposed itself as unhinged. The International Monetary Fund warned in September 2012 that the global economy was just as much at risk of a crash now as it was before the last one. It urged global regulators to dive much deeper: to consider banning banks outright from certain of their continuingly risky activities, instead of simply hoping that tougher capital requirements will guarantee financial stability. It urged them to monitor 'non-banks' such as hedge funds, checking for risky practices. It advocated other obvious safeguards that the uninitiated might think would have long since been put in place.

In October 2012, the Bank of England's director for financial stability, Andy Haldane, tabled similar fears. Banks deemed 'too big to fail' at the time of the bailout in 2008 continued to grow, he warned, concentrating capital in a way that, far from dismantling risk, worsens it.

As 2012 ended, a further blizzard of Libor revelations hit the press. It was the turn of UBS to stand in the spotlight, not to mention a record 1.5 billion dollar fine. One internal e-mail from a trader keen to do a little fixing gave a flavour of the prevailing culture particularly well. 'I will fucking do one humongous deal with you . . . Like a 50,000-buck deal, whatever . . . I need you to keep it as low as possible . . . If you do that . . . I'll pay you, you know, $50,000, $100,000 . . . whatever you want.'[5]

People like me who know a little of what goes on at banks found it impossible to read such words and not conclude that they spoke of a culture of permissiveness amounting to senior collusion. People who work in these places know that e-mails should be written as though they are open documents within the institution.

In January 2013, the FSA and the Serious Fraud Office launched an investigation into Barclays over their 2008 Qatar bailout, suspecting that they had loaned Qatar the money with which to buy the shares to invest in the bank and save it from the clutches of the UK taxpayer. All this had been going on while I whacked golf balls round a Singapore country club with Bob Diamond, Phil Mickelson and others. Diamond now faced trial for fraud. He was not going to be alone, it seemed.

Meanwhile, the consequences of the 2008 financial crash continued to unfold. In anguished Greece, the prime minister professed that democracy itself was now under threat. In Davos, at the annual World Economic Forum gathering in January, the mood was one of widespread concern. Although no other eurozone country had yet to crash and burn quite as badly as Greece had, many delegates wondered whether the best we could hope for now was a world of extended downturn.

Davos likes to discuss what the next big idea for the global economy might be. Fracking was aired as a candidate in January 2013. Summarising her week with Davos Man, Heather Stewart of the *Observer* wrote: 'Governments are increasingly pinning their hopes on [shale gas] as a low-cost energy source, releasing them from dependence on the volatile Middle East.'

***

Climate, meanwhile, had found its way back onto the global agenda. It took an assault on New York by a megastorm to do it. On 30 October 2012, Hurricane Sandy swept a thirteen foot storm surge into Manhattan, flooded the subway and shut the New York Stock Exchange, among other things.

America was heading for the polls in a presidential election. Mitt Romney had made the big mistake of playing climate change for laughs with his tribe. *Bloomberg Businessweek*, announcing support for Obama, ran a spectacular front cover on 1 November. 'It's global warming, stupid!' the headline shrieked.

This rather broke the mould of media coverage of climate change in the United States since Copenhagen. It may not have won Obama the election, but it hardly helped Romney. Climate change had not been mentioned in any of the three presidential TV debates.

On 14 November, the re-elected President addressed climate change in his first press conference. On 21 January, in his inauguration speech, he majored on it. By February, he was issuing Congress with an ultimatum: back me on climate or I will go it alone.

He had an early chance to face down the oil and gas lobby on climate. On 7 January, 70 US NGOs urged the President to deliver on climate in his second term. They had colluded in a strategy of climate silence in his first term, but this, they said, was now firmly over. On 17 February, the largest protest in US history put pressure on Obama to reject the Keystone XL tar sands pipeline. Many Americans now clearly see this as the defining battle for his legacy. Whether he will pass the test is still not clear at the time of writing. Bill McKibben, 350.org and the rest of the anti-Keystone campaign will have a lot to do with whether he does. In October 2012, McKibben had kicked off a new campaign, based on Carbon Tracker's first report on the carbon bubble. He called it 'Do the Math'. In November, he embarked on a national lecture tour to whip up support for the moral case for divesting in fossil fuels.

The climate science added nothing but urgency to the case. 2012 was in the top ten warmest years ever. Each twenty-first century year has now been in the top 14. The US National Centre for Atmospheric Research has found that the computer-based climate models with the highest projected temperatures are the best at replicating real climatic data, suggesting that the worst-case estimates by climate scientists of warming are more likely to be accurate than the best-case estimates. My website chronicles other such depressing discoveries.

Another full IPCC scientific assessment report is clearly overdue. The next, the fifth, is to be published in 2013, after a gap of fully seven years. It 'will scare the wits out of everyone' and 'shock nations into action', says the former chair of the Framework Convention on Climate Change, Yvo de Boer, now at KPMG.

But on the other hand, the myth makers will be hard at work providing a counter-narrative. In February, it emerged that a vast network of climate denial organisations has been built up by American conservatives using a secretive fund with the anodyne name Donors Trust. This body funnelled nearly $30 million to denial organisations in 2010. ExxonMobil and the conservative Koch brother oil billionaires channelled some $5 million between them.

Against the tide of disinformation, and worse, that this machine would create, Carbon Tracker and other organisations trying to alert the world to the carbon bubble ploughed on with efforts to awaken the capital markets to the risk of stranded assets. In March, the Carbon Tracker team commissioned a report from the ratings agency Standard and Poor's. The carbon bubble means that the business models of tar sands companies are at risk, it concluded.

The report provides a perfect reason, rooted in the world of financial risk, to say no to the Keystone XL tar pipeline.

Stirling Square, London, 5 March 2013

In one of the palatial buildings owned by the beleaguered investment bank now calling itself Citi sit a hundred investment bankers and their invited guests. We are here to discuss what our hosts, Citi and the International Finance Corporation (IFC), refer to as 'real-world solutions for bridging the climate investment gap'. The head of treasury at the IFC explains what they mean by that.

Five years ago the IFC was investing two-thirds of its capital in fossil fuels, he explains. Now it is investing two-thirds in renewables. We raised our first billion dollar green bond recently. So we are making some progress. However, we are merely scratching the surface. If the world community is to keep the global thermostat below the two-degree danger threshold, business will need to mobilise a thousand billion dollars of investment for clean energy this decade. That is what we mean by the climate investment gap.

Failure to close the gap is inconceivable, he continues. As the IFC's parent, the World Bank, has shown recently, he explains, if we continue to capitalise fossil fuels instead of clean energy, we are currently on a course that risks destroying global food and water supply.

The starkness of this analysis impresses me, on one level. I have never heard anyone, let alone a financial bureaucrat, reduce the climate change risk to such a bare sound bite.

Next, two economists are asked to set the scene for our discussion by describing their current projections for the decades to come.

The chief economist at Citi describes a global economy wherein GDP and energy consumption go on rising essentially without constraint, either regulatory or from resource availability.

BP's head of energy economics is next up, to show BP's latest energy forecast. A new era of fossil fuels leaps onto the screen, aesthetically decorated with BP's green and yellow sunflower logo. The crucial pie chart shows almost half the world's energy, 36% higher in 2030 than today, coming from oil and gas. Another shows that the hike in global oil production needed to fit that prescription would be coming entirely from unconventional oil.

It is almost as though this man hasn't heard a word of what the IFC executive was saying.

I am on the discussion panel that follows. They have asked me to talk about the capital needs of the solar revolution. But what I want to talk about first is that damn BP forecast.

It isn't a forecast, I say, when my turn comes. It's a suicide pact.

I look at the BP man, now sitting in the front row.

You say that forecast is just a forecast, that it is not necessarily what BP wants. I have been invited to speak as a frank friend today. Let me now do so. I feel – I'm sorry, don't take this personally – that you risk being disingenuous. BP's lobbyists are pulling out all the stops to try and make that forecast *happen in reality*.

Smiles spring to faces in maybe 10% of the audience. It seems that some investment bankers like a little frankfriendedness. Either that or the prospect of a fist fight.

In the tea break that follows I get a lot of what I am used to by now.

I'm so glad you said that. Someone needed to.

Yes, but why don't a few others speak up?

We return to the discussion. We hear passionate advocacy of climate bonds. This is the key, says green bonds guru Sean Kidney. This is how we can unlock the passive, supposedly risk-averse but actually risk-additive trillions of the pensions funds. All we need are the processes to do it.

We hear articulate campaigning for energy efficiency. This is mission difficult, not mission impossible, says green infrastructure banker Jonathan Maxwell. This is how big institutions save big money, never mind the environmental and social upsides.

Then the man from Blackrock. It's all great stuff, he says, and necessary as hell. But it's going to take policy to make it happen.

Here we have it again, I think. The endless game of responsibility switching. For the energy and finance incumbencies, the policymakers have to act first. For the policymakers, the incumbencies have to act first, voluntarily. And stop their default, below-the-radar, mass lobbying to protect the status quo and hustle up easy bonuses.

Elsewhere in Citi, I know – here in London and on Wall Street – the shale gas and tight oil gravy train barrels on. Many more Citi bankers are attending those meetings than this one.

In October 2012, one of Saudi Arabia's top spokesmen, Prince Turki Al Faisal Al Saud, told the Global Economic Symposium in Brazil that he hoped the kingdom might be powered entirely by renewable energy and other low-carbon forms of energy within his lifetime, so that oil production could be preserved for export. The reason for this enthusiasm for solar and other renewables we have encountered already, notably in the report by Citi in September.

A statement such as this tells us two hugely important things. The first is the window it gives us into Saudi thinking about oil. If they really have so much crude oil left to extract, would they be worrying to this degree? Could they not simply open up new giant fields?

The second is the credibility given to renewables and their ability to run the electricity requirements of modern economies. It is not difficult to imagine the Saudi government watching the dire progress of nuclear elsewhere in the world, conflating all the industry's setbacks with the problems a huge nuclear programme in the desert nation would pose them in terms of cooling water, and concluding that nuclear power was in reality not much of a runner. And in terms of renewables, Saudi Arabia has a limited wind resource. So Prince Turki's thoughts must have a major emphasis on solar. Certainly the fact that the Saudi government has announced a $100 billion investment programme for solar suggests so.

Very few governments have talked about renewables 100% powering national economies. So the question arises as to whether such a feat is feasible. A senior Saudi prince would probably not tell the world that he and his government think it is unless significant others were saying the same, at least in private. In Part II of the book I shall argue that they would be justified in this view.

Others have been demonstrating in unilateral action that they think renewables might have a pivotal role to play in powering economies in the years ahead. In October 2012, furniture giant IKEA announced that it planned to be entirely self-sufficient in renewable energy, across the entirety of its global operations, by 2020. This unprecedented programme entails 300 stores in 26 countries and investment of $1.9bn in wind and solar projects by 2015. In January 2013, Warren Buffett showed that he might have awoken to the potential by buying half a gigawatt of solar PV farms. My website log gives other such examples.

Outside the corporate world, in the citizenry, we find many other examples of positive action on clean energy. By the end of 2012, over 600 clean-energy co-operatives had been set up in Germany, many of them revolving around wind and solar. Proliferating at the rate of one every two days, these co-ops are becoming larger, more professional and more closely linked to cities.

All this suggested that an explosion of financial innovation might not be an unreasonable supposition, and that much of it might involve the same kind of people power manifesting itself in the German energy co-ops. Entering 2013, the first stages of a potential megatrend in crowdfunding and peer-to-peer lending were taking shape in the world. A June 2012 report by Bloomberg showed that crowdfunding has to date provided more than $500 million by aggregating micro-investments from many thousands of motivated and unaccredited individual investors, and suggested that if just 1% of current retail investment in savings accounts, money markets and US Treasuries could be diverted to crowdfunding – a perfectly feasible suggestion – $90 billion would become available.

Growing numbers of people were deserting the big banks in anger, in favour of ethical banks, credit unions and the like. On 17 December 2012, the Bank of England's Andy Haldane drew an extraordinary conclusion. 'The mono-banking culture we have had since the 1990s is on its way out', he told the *Independent*. 'Instead, we are seeing a much more diverse ecosystem emerging with the growth of new non-bank groups offering peer-to peer lending and crowdfunding which are operating directly with a wider public.'[6] One might expect to hear a visionary green thinker talk about banks being disintermediated by people power, but an executive director of the Bank of England? Two months earlier he had also conceded in a speech that the Occupy movement was correct in their analysis of what caused the financial crash of 2008.

At the time of writing, the incumbency myth machine is in overdrive trying to do down renewables, and the notion of investing in them. In February 2013, Fox News provided a classic. Solar energy is heading nowhere in the United States, said an enthusiastic business correspondent, and is only working in Germany because that country is sunnier than the States. This made amusing viewing on YouTube when juxtaposed with the insolation map showing that nowhere in Germany is more sunny than the least sunny area of the USA, the Seattle region. But once again, far more people would have believed the myth than saw the truth.

Nonetheless, there are days when it seems the momentum behind renewables offers genuine encouragement. Often such days are tinged with irony, for me. One such was 28 February when, four years after quitting solar energy, Shell announced that solar was destined to become the world's biggest energy source.

Excel Exhibition Centre, Docklands, London, 6 March 2013

The annual Ecobuild trade fair. The international construction industry is crammed into two exhibition halls, each one capable of housing the full fleet of an international airline. People have piled into these hangers in their tens of thousands. They wander for miles along row after row of stands decked with every kind of product on offer to the construction industry. Everything to be seen is designed to cut or eliminate environmental damage in some way.

I have been going to these shows for more than a decade. When I started, you could get round in a few minutes. Now it takes hours. There are far more suits today, far fewer check shirts.

The clean-energy industries are here in force. Energy efficiency performance is a main selling point for every brick, valve, heating system and window in sight. Renewables companies from all around the world vie with each other in the size and glamour of their stands.

Solar is no exception. The Solarcentury and SolarAid stands, back to back in the heart of solartown, are heaving with people being shown solar at work in real projects on all scales from utility-scale power plants to tiny solar home lights. I note with pride that other solar companies have provided volunteers to join Solarcentury staff in volunteering on the SolarAid stand.

I join a lunchtime discussion panel in the conference run in parallel with the exhibition. This too is packed. The title of the session has taken a leaf out of the *FT*'s book in portraying the problems of the renewables industries in the face of the new gas and oil narrative. The theme: 'Is the micro-generation boom over?'

No, I say. Take a walk around.

But I have also been to oil industry trade shows. At one, in Houston, there was a whole car park just for the helicopters. I know the massive inertia we are up against more than most at Ecobuild. And it is not just the battle between the incumbency and the 'insurgents'. We have our own civil wars. This year there are far fewer Chinese participants at Ecobuild. The EU–China and US–China solar trade disputes are taking their toll.

But every year this spectacle lifts my morale. I leave the show with renewed hope in my heart that this is the stuff of a road to renaissance.

Cautious, qualified, caveated. But hope, nonetheless.

Bloomberg UK HQ, City of London, 18 April 2013

In a glitzy and cavernous data centre in the heart of the capital markets, hundreds of analysts and journalists sit at futuristic terminals totting up the scores in the miriad indices and benchmarks of modern capitalism. As I make my way through security, I look at them all through the plate glass, worker bees in a honeycomb of computer screens, and wonder how many are keeping the score with climate change: whether any of them have an estimate for the year when our overheating atmosphere begins to destroy wealth faster than the markets they analyse can create it. Atmospheric carbon dioxide concentrations will exceed 400 parts per million within a few weeks.

I am here for the launch of Carbon Tracker's second report. The Bloomberg auditorium is overflowing. Our latest conclusions focus on the capital that the energy incumbency is lining up with which to fuel the global suicide pact over the next decade, or, to use Carbon Tracker language, to risk stranding. It is a joint production with Lord Nick Stern's Grantham Research Institute on Climate Change at the London School of Economics. The Carbon Tracker team is beginning to accrete some serious credibility.

Lord Stern himself kicks off. The former Treasury chief economist seems to become angrier about the international climate-policy impasse with every presentation he gives.

You cannot believe two things simultaneously, he tells the audience and the cameras. You cannot believe that we have a reasonable chance of holding to two degrees *and* believe that hydrocarbon companies are anywhere near valued at the right price.

James Leaton, Carbon Tracker's star analyst, elaborates on this evaluation dilemma as he relays the reports findings. His delivery, as ever, is measured and thoughtful.

On the *Guardian*'s website, Bill McKibben and I have used somewhat more emotional language. Suppose you weren't worried that we humans are destroying our water supply and eroding our ability to feed ourselves by burning coal and gas and oil and hence changing climate, we posit. Suppose you thought that was all liberal hooey. What might worry you about fossil fuels instead? How about a six-trillion-dollar bet, including a big slug of your own money, on people not doing what they have said they are going to do, and that some have already sworn to do in law?

Six trillion dollars is what oil, gas and coal companies will invest over the next ten years on turning fossil-fuel deposits into reserves, assuming last year's level of investment stays the same.[7]

After the presentation of the Carbon Tracker results, I chair a panel discussion of financial experts who have agreed to give their reactions to the new findings. I have met none of these people before today.

Paul Spedding, Oil and Gas Analyst at HSBC, decides this is the day to tread a new frontier. What struck me about the report, he says, was the sheer scale of the reserves within listed oil that might be unburnable. Potentially only 20% of fossil-fuel resources are burnable, leaving 80% as a potential carbon bubble. I suspect most investors would put that split the other way around, so this is not an issue that is reflected in current valuations in my view.

I listen to Spedding in surprise as he agrees with Lord Stern. I had not expected this frank a reaction from such a senior player representing an institution in the heart of the incumbency.

If we are serious about addressing climate change, he continues, there needs to be a shift in the business models for fossil-fuel companies, nearly all of which are pursuing an 'invest to grow' strategy. This is reflected in the fact that most companies have steady-to-rising capital budgets. Failure to address the threat of unburnable carbon could leave players exposed to material asset write-downs and wasted investment, both potentially destroying shareholder value.

This is dynamite, I think to myself. If we can persuade a few other mainstream oil and gas analysts to start talking this way, we are going to have a real chance of impacting the oil, gas and coal industries' financial licence. Regulators would feel pressure to push through requirements that the risk of asset-stranding be recognised. All sectors of the financial chain would feel pressure to recognise that risk independent of regulatory requirement: to jump, as it were, ahead of being pushed. With the risk being more and more discussed, more and more capital would eschew carbon.

350.org is already beginning to persuade universities and cities to sign up for divestment in fossil fuels. With citizen pressure using 350.org's language of morality, and institutional pressure using Carbon Tracker's language of capital, we stand to create a pincer movement.

Paul Spedding continues. The least companies should be doing is disclosing more details of their resource base, he explains. At present, statutory filings from oil companies only give details of proven reserves. But this only represents around a quarter of their overall resource base. Probable reserves account for a further quarter, with resources making up the remaining half. For shareholders to be able to assess the risks to their investments, oil companies should give far greater disclosure. In the longer term, company boards need to decide whether the business-as-usual, invest-to-grow model is sustainable. After all, the central conclusion of this report from Carbon Tracker, and from various IEA studies, is that growth in energy demand must slow. Coal and oil demand will need to fall and the only fossil fuel that can continue to grow, albeit slowly, is natural gas.

Spedding comes to his astonishing conclusion. To survive in such a world, he says, energy companies must refocus and cut capital expenditure, instead returning capital to shareholders in the form of dividends and buybacks. Such a model has been embraced by some companies, such as Conoco. Known as 'shrink-to-grow', it was implemented not for environmental reasons, but to boost shareholder value. With the risk of value destruction from an 'invest-to-grow' strategy, companies need to focus on per share growth rather than absolute growth. From a shareholder perspective, it should not matter whether growth is achieved from buybacks or from investment. From an environmental perspective, buybacks or additional dividends are clearly preferable to investment in what may well turn out to be stranded assets.

After Spedding's portentous message for the oil industry, a senior executive from Moody's, the ratings agency, turns the spotlight on coal. Investors should and will be paying particular attention to the stranded assets of coal companies, says Steve Oman, Senior Vice President for Corporate Finance, because coal is clearly the greater villain. The report is invaluable, he adds. I applaud the points on the need for disclosure.

Howard Pearce, Head of Pension Fund Management at the Environment Agency Pension Fund, also elects to speak out.

I manage a £2 billion pension fund, he says, and this is my view. The Carbon Tracker report should be read by every chairman and every chief investment officer of every UK pension fund. Unburnable carbon is not an asset, it's actually a liability. That doesn't really seem to be priced in or thought about by many people at the moment.

On the stage, chairing the discussion as these mainstream capital-markets players deliver their seismic conclusions and the audience questions them, I have to work hard not to let my excitement show. Facing a camera for a YouTube summary of the event, I can't conceal it.

We're fascinated to see how the Bank of England reacts to our report, I say. They are responsible for the stability of the capital markets. And right now they are allowing the carbon bubble to go on inflating, with nothing done. I think that has got to change.[8]

I leave the event at Bloomberg as excited as I have ever been about the possibility of a turning of the tide on climate change. I know some philanthropic foundations share this hope for the work of Carbon Tracker, and that six-figure cheques are in the offing for funding of future work and campaigning.

Could this be the best chance to start switching capital from carbon to clean energy on the scale needed, I ask myself? And how will the other systemic risks play out while this part of the narrative unfolds? Might there be a road to renaissance in their interplay?

Or is it actually too naive, even too late, to hope for such an outcome?

Part II

A future

Chapter 18

What next?

The anatomy of the biggest crash

Those who try to predict the future in our complex world risk looking ridiculous when hindsight turns its searchlights on them. Forecasting how the five systemic risks I have described will play out individually is hard enough. Attempting to foresee how they will interplay involves a wholly different level of hazard. But let me offer my best guess, and try to follow a logic chain in the process.

Part I has chronicled a surprising profusion of players predicting an oil crunch within a few years, from diverse sectors spanning academia, industry, the military and the oil industry itself, including until recently the International Energy Agency or, at least, key individuals or factions therein. As we have seen, many predict crisis within a window from 2015 to 2020.

Those forecasting a crash by 2015 at the latest include, notably, a taskforce of British companies across multiple sectors of industry and the US armed forces. Both these predictions were made in reports published in 2010, since which time a blizzard of hype has emerged from the oil industry and those who capitalise it about the potential for tight oil and shale gas production in America. So the first question to ask is whether those more immediate warnings of crash potential still hold.

The soaring drilling rates in America have succeeded to date in lifting tight oil production by something approaching two million barrels a day. It has taken more than half the world's non-FSU non-China oil and gas drill rigs to do that. The world is consuming around 90 million barrels a day and depleting easy-to-access crude by over four million barrels a day of capacity each year, a figure that could easily accelerate, as the IEA has warned. Though the tight oil additions are welcome for the US economy, they can hardly be material in the global picture, even if they can be maintained.

They are much more likely, given the drill rig effort required, to turn out to be the Red Queen's Race many analysts suspect them to be.

As for US shale gas production, gas contributes well under 1% of US transport fuel, and some of the shale gas-fields are widely suspected of being already in decline, even if drilling could be maintained at less than the marginal cost of production. But that is a theoretical question anyway, because so many drilling rigs are switching to more lucrative shale oil production. Shale gas drilling has dropped off a cliff since 2009. It is only a matter of time now before US shale gas production falls. This is not material to the timing of a global oil crisis.

So I stick with the prediction of an oil crash by 2015 at the latest, absent renewed financial crisis cutting demand in the interim.

A global oil crash would inevitably trigger a financial crash in its wake. The highest ever oil prices helped trigger the financial crash of 2008, as high gasoline prices turned a stream of American mortgage defaults into a flood. We should expect no different should oil break price records again so relatively soon. Modern economies are ill prepared to withstand high oil prices, and failing economies do not generate the income indebted governments need to service their debts.

If many in the financial commentariat are to be believed, a second financial crash is a high risk imminently even without the trigger of a high oil price. In that scenario, the weight of debt that we have allowed to accumulate around the world will prove just too heavy for the financial system. As things stand, a seemingly small event holds the potential to trigger the mass failure of banks. The ongoing euro crisis provides perhaps one possible tripwire. Banks have enormous bond holdings of governments in whose economies trauma is building as I write, simply as a result of indebtedness and the shakeout of the 2008 crash.

There are other potential triggers. For example, the Bank of England has warned that a private equity crash could kick off the next wave of crisis. Debt-laden companies bought by private equity firms in the boom years before 2007 will have to refinance huge loans in 2014, in much tightened credit conditions. What happens if many of them fail? This is the kind of thing that can crash a system as fragile as the capital markets have become.

History shows that oil demand drops in the global economic downturn following a financial crisis, releasing pressure on tight oil markets. But in a recessionary world today, or even a global depression, how long would that demand pressure dissipate for? The demand for oil in China, India and the major oil-producing countries is likely to be enduring. Think of the soaring domestic oil consumption for infrastructure construction in oil-producing countries, notably Saudi Arabia; the advent of very cheap cars in India; Chinese airport construction plans; the consequent fall in oil exported since 2005, and so on. Then think of that plunging crude oil depletion curve in the IEA's 2008 *World Energy Outlook*, how the IEA says it could steepen, how the number crunchers in the Association for the Study of Peak Oil (ASPO) say yes it could steepen, but the line as forecast by the IEA isn't steep enough to begin with. Think how IEA whistleblowers have endorsed the ASPO view, saying that political pressure from the US stopped the full story being told. Superimpose on that the vast expansion of production from difficult-to-access deep-water crude or unconventional oil that would have to substitute for that depletion of crude oil. Looking at all this, it is probable that a global oil crisis could only be delayed for a short period by renewed financial crisis.

How might the other three systemic risks discussed in this book – climate change, carbon bubble and shale surprise – play out in parallel?

The economic and environmental harm that we stoke with ongoing greenhouse-gas emissions builds in small increments. Ultimately it is the cumulative harm to the climate system from the increments that poses the greatest threat of all the five systemic risks I consider. But I suspect the climate crisis is less likely to trigger a full-blown system crash in the next few years. In the current decade, I suggest, world events are more likely to be driven by crashes in the oil and financial sectors.

This could easily be wrong, I should add. There will be individual reinsurers who suspect an insurance crash might prove me incorrect, and some food experts who suspect a food crisis might.

As for the risk that assets stranded in the carbon bubble will trigger a crisis, it is not impossible to envisage the rush of panicky climate policy-making in the face of serial climate disasters that would trigger enough stranding to crash the system. But is such a sudden reconfiguring of risk perception likely to materialise in the next few years? Based on the last quarter-century history of the inability of world leaders to take firm action to cut emissions – even in the face of the increasingly dire recent evidence of a climate already going awry – it seems improbable. It is more likely that the bubble will go on inflating, deepening the problem down the road.

Of course, if a degree of risk-recognition about asset-stranding comes into play, in the way Carbon Tracker and others seek, the bubble might not inflate further. Flight of capital from carbon fuels might even begin deflating it. So much the better for the prospects of renaissance if so.

The risk of a shale crash may also be working on a longer time frame. Oil and gas companies have a significant general ability to sell assets to cover strains on cash. This is how Chesapeake has held off its balance sheet problems without going bankrupt so far, for example. And as for the exportability of the shale narrative beyond US shores, the denouement will not play out for some years. Analysts have told the *Wall Street Journal* that it could take as long as ten years to export the US shale boom. In the UK, for example, the first commercial shale gas well is not scheduled until 2014 at the earliest. Each and every well will face intense local political pressure before it can go ahead. Other systemic risks will play out before this drama reaches its full crash potential.

In summary, if my suppositions are correct so far – a big if – the world faces either an oil shock by the end of 2015 and a financial crash soon after, or a second financial crash before then, followed by a delayed oil crash some years later.

***

We can think of these two scenarios as 'Oil-Fails-First' and 'Finance-Fails-First', for ease of reference.

Six big questions then arise.

To what extent could we hope to head off or soften the blow of an Oil-Fails-First scenario?

To what extent could we hope to head off or soften the blow of a Finance-Fails-First scenario?

How much time would we have in the Finance-Fails-First scenario between the financial crash and the oil crash, and would it be enough time in which to soften the blow of an oil crash?

Which is the more likely of the two scenarios?

What would be the implications of an oil crash?

And then finally, and absolutely crucially, what are our chances of rebuilding society beyond the crisis or crises, and to what extent, given the other systemic risks we are taking? And relatedly, what could that society look like?

The sixth question is the basis for my last chapter. Let me now consider the other five in turn.

Question One: To what extent could we hope to head off or soften the blow of an Oil-Fails-First scenario?

If the prediction that there will be a crash in global oil supply by 2015 at the latest is correct, we would have a maximum of 32 months from the time of writing to mobilise enough alternatives to oil-based transportation to dismantle the potential for a crash. I know of nobody who thinks this would be remotely possible, even if the political will were there to begin mobilising tomorrow. Energy demand-management campaigns can achieve a lot, as the Japanese post-Fukushima effort has shown. No doubt transportation campaigns could achieve analogous results. But modern renewable energy is not yet 10% of the global energy mix, and renewable and non-carbon fuels are an even tinier fraction of the transport mix.[1] There would be too much of a gap to make up. We would face the full shock factor of the third great global oil crisis.

Question Two: To what extent could we hope to head off or soften the blow of a Finance-Fails-First scenario?

Here there is much more we could do. The history in Part I gives a flavour of the proposals that abound for deconstruction of the problem via fundamental reform of the financial sector. More feature on my website. Our collective problem, however, is that recent history also shows we do not have leaders capable of championing them and executing them. It seems most likely that the capital markets will bumble on in their mire of greed-based hard-wired short-termism until the next crash. As for what will trigger it, we are spoilt for choice.

Question Three: How much time would we have in the Finance-Fails-First scenario between the financial crash and the oil crash, and would it be enough time in which to soften the blow of an oil crash?

In the countries fundamental to the answer, those with the fastest-growing oil demand today – China, India and the major oil-producing countries – the economic downturn after the 2008 financial crash was less marked than in the main industrial economies. In part this resulted from lower exposure to toxicity in their financial systems. It also happened because of the intensity of the societal pressures to keep economies growing, and so burning oil. The leaderships in China and the Gulf oil states have to keep their burgeoning middle classes content. If they don't, they face existential threat. This is not just about being voted out of office for four years, as it is for political parties in democracies. It is about the end of systems and dynasties. The Chinese Communist Party could not expect to survive a revolution. Neither could the House of Saud. The rest of the global economy might be in a dire state, but the vast infrastructure programmes in those economies would have to be kept going, and they would be just as oil intensive post-crash as they are today. On top of this, the democratically elected governments in high-oil-consumption countries, led by India, would be trying no less hard to keep their electorates content enough to stay in power.

Looking at this picture, it seems probable that the global oil crisis could only be delayed by a few years. That would be time to make some material policy interventions aiming to accelerate clean energy and soften the blow of an oil crash, though not to head it off. But this, like the prospect of heading off the financial crash, is probably a theoretical consideration. Given the evidence in Part I of how easy it proved for the incumbency to suppress the perception of the peak-oil threat after the 2008 financial crash, it is difficult to imagine a different pattern after a second crash.

Question Four: Which of the Oil-Fails-First and Finance-Fails-First scenarios is the more likely?

The answer to this has to be rooted in the day-by-day monitoring of the real fallout from the 2008 financial crash and appraisal of the build-up to the putative oil crisis. On that basis, I find it increasingly difficult to imagine how the financial system can keep going until the end of 2015 without another crash. The wall of private-equity debt needing refinancing in 2014 has the whiff of straw and camel's back about it, as the Bank of England has suggested. Then there are the debt-drenched vagaries of the rolling eurozone crisis. I suspect the Finance-Fails-First scenario is the better bet.

If that is correct, and if I am correct in the assumptions about enduring oil-demand pressures in China, India and the main oil-producing countries, then we face a few years of global recession or depression before having to find out in reality what the answer is to the next question, which is a big one indeed.

***

Question Five: What would be the implications of an oil crash?

The history of the credit crunch and financial crash suggest some analogues for how the peak-oil crash might play out. In 2007, it was astonishing to watch how quickly the financial services industry came to realise it had been guilty of buying in to a rootless belief system. One week the incumbency was chanting mantras about scaremongers not understanding the sophistication of investment bankers in excising risk from mortgage-backed securities. Then came enforced acceptance that perhaps a few mortgage-backed securities might indeed be toxic. Within weeks of that, almost everyone had come to accept that a vast slug of the asset class might be toxic. Suspecting other banks of being massively exposed, just like they mostly were themselves, banks stopped lending to each other and, with frozen credit markets, slow strangulation of economies began. It took just over a year for the full unpleasantness to work through to the point of the crash.

We now know the broad psychology of what went on. Relatively few players in any one bank really understood the detail of complex derivatives. Vast numbers of players in the banks, and down the product chain, took it all on trust. The system simply couldn't believe it was fallible, and certainly not on the scale it proved to be.

This is how it would play out with peak oil, I suggest. Only a few players in any one oil company, or any one government, are charged with a complete audit of their own reserves, and all the data pertinent to that overview. And they are only party to their one company's or government's data. Even if they genuinely don't think they have a problem, then they can't know for sure about the regional picture, much less the global picture. They simply take it all on enculturated trust. The OPEC countries report to OPEC headquarters the reserves they want to see posted. The OPEC staff tot up the total reserves. The BP Statistical Review of World Energy regurgitates the OPEC reserves data annually, with only a tiny footnote health warning, and BP bosses pretend every year that they have produced the most authoritative document possible. Journalists, policymakers, business schools and just about everyone else echo all this without too many questions. And so the belief system is perpetuated.

Peak oil itself will not trigger the crash. It will be the perception of peak that will trigger the crash. Peak oil will only be visible in the rear-view mirror, after some years of falling production. Even then some people will question whether we have in fact experienced peak oil, saying the industry can lift production above the supposed peak again if only it were just left alone to do so. But let us envisage a change of perception akin to the first wave of defaults on sub-prime mortgages. A giant Saudi oilfield suffers a collapse in production, say. Or an entire nation's oil supply is shut off by a sudden 'above ground' issue. The oil price rises to crisis point.

The calls begin for spare capacity to be brought onstream, as we have seen happen recurrently in the history in Part I. The Saudis say they are endeavouring to do that. But their production in fact falls in the next months, and their exports fall still further because domestic consumption is still soaring, even with all those gigawatts of solar they will have installed in the interim. More and more people come to think the Saudis do not have functional spare capacity any longer. The concern spreads with the brushfire speed it did in the capital markets in the run up to the credit crunch.

Those not following play in the oil markets, beyond the plentiful and constantly repeated mantras of the incumbency, now look to the aspiring 'Saudi America' to lift production faster, or to the taps in the tar sands opening wider, or to tight oil to be produced in countries yet to produce any, or to Russia to stop holding its neighbours to ransom, or to Brazil to hurry up and produce all that deep-water oil, or to gas to be rendered liquid faster, or to coal to be turned into liquids at scale. They find themselves disappointed.

Many who are following play in detail are not surprised at all. There may be trillions of barrels of oil in these 'resources', but there surely aren't taps that can be opened much wider at will and in haste. Especially when such a huge proportion of the world's drilling rigs are in North Dakota and Texas, fracking small quantities of oil out of tight wells in which production tends to drop off quickly.

Bang. Panic.

The oil price soars. Markets crash. This is the biggest of them all, the Bank of England says. We didn't see this one coming.

We did, the 'peakists' might suggest, risking the ire that descends on people who say 'we told you so'. Why not indict the oilmen who foisted this on the world with their recklessness?

Ridiculous, BP and the rest will say. There is plenty of oil. It was above-ground factors that caused this scarcity. Peak oil is nonsense.

No, the peakists will respond, it was the combination of below-ground factors and above-ground factors that did it. Peak oil is peak oil – the global peak of production, however caused. We have passed it. You know this now, and are obfuscating as desperately as you do after oil spills.

Only lawyers scenting deep-pocket blood will care about this bickering, because the crisis will be worsening rapidly. Within the oil-producing countries themselves, manic post-mortems will be under way. Could it be that we have been deluding ourselves about how much affordably accessible oil we have left, in the same way so many financial-sector players deluded themselves about mortgage-backed security assets?

We may well have, comes the increasingly common response. Pre-cautionary movements emerge in multiple governments. Husband oil for domestic use, they say. Slow the exports. In the case of less important governments, stop exporting oil at all.

We have seen hints of this potential in Part I, in the comments of the Saudi king, and in doubts expressed about reserves in the Kuwaiti parliament.

And so, just as the UK Industry Taskforce feared the worst case to be, tankers stop arriving in some oil-importing nations, or are diverted on the high seas to bidders offering even more extortionate prices.

Now we have utter panic.

Events in the UK in the autumn of 2000 point to the kinds of stresses that would have to be endured. Truckers aggrieved at high diesel prices at that time elected to blockade refineries in protest. By the third week of the so-called 'fuel crisis' that resulted, supermarkets were rationing bread and the government of the day was poised to deploy elements of the army as a reserve oil-delivery force.[2]

There would be a big and daunting difference this time, however. In that episode, the shortages of fuel were artificial and temporary. The crisis feared by those who worry about premature peak oil is of an altogether different nature: systemic and inescapably permanent, so long as economies remain anything like as oil dependent as they are today.

If the analysis is correct to this point, we will have arrived irredeemably in a time of consequences.

Chapter 19

The power of context

Energy and security

When armed forces turn their minds to the strategic implications of oil production, as we have seen in Part I, they have tended to side with those worrying about early peak oil. The 2010 Bundeswehr study of peak oil warned that the consequences of premature peak oil could include 'economic collapse'. The 2010 Joint Operating Environment Report of the US Joint Forces Command warned of global depression and the potential rise of totalitarian regimes seeking 'economic prosperity for their nations by ruthless conquest'.

Consideration of oil security in the context of the military's own energy requirements certainly explains much of their concern, and offers a sobering window into the strictures of the coming crisis. In 2011, the average US soldier consumed 22 gallons of gasoline per day, and the Pentagon spent over $17 billion on fuel. Every $10 increase in the price of a barrel of oil costs America's military well over a billion dollars. And of course the issue goes well beyond the monetary cost. A Marine Corps evaluation found that fuel and water convoys have accounted for 10% of casualties in Iraq and Afghanistan.

Overlaid on this is the question of the root causes of terrorism. As a recently retired Marine Corps general put it: 'America sends nearly one billion dollars per day overseas to purchase foreign oil. It is undeniable that some of that money ends up in the hands of groups that wish to do us harm. We cannot afford to be in the position of funding both sides of the war against terrorism, from either a security or an economic standpoint.'[1]

As for the solution, a May 2009 report by the Military Advisory Board of the Center for Naval Analysis, a panel of some of America's highest ranking retired admirals and generals, concludes that there is no possibility of drilling a way out of the oil problem, and calls for a 30% reduction in overall American oil consumption, to be achieved by using energy more efficiently, while developing and deploying renewables and alternative fuels.[2] In essence, this military panel is calling for the same general suite of policy responses to the oil depletion threat that the UK Industry Taskforce on Peak Oil and Energy Security has called for.

What of the thinking that militaries do that never sees the light of public scrutiny? Let us take a big swallow and ask ourselves to what extent the kind of posturing James Schlesinger revealed in Part I might be at work behind closed doors in the Pentagon, in NATO and their equivalents. Do the Americans have plans drawn up to grab the Saudi oilfields by force, if circumstances require? Circumstances that might include the Saudis shutting off exports before the 'Saudi America' unconventional oil scenario reaches full development?

Let me hazard a guess. This is the most powerful military that the world has ever seen, in relative per capita terms. I would not be surprised. Filed under 'Options for the President', of course.

Do the Chinese? With their one cast-off Russian aircraft carrier, and a military budget one sixth the size of America's, I would have to doubt it. They would need to advance their strategic goals by other means.

Do the Russians? Maybe I wouldn't worry in Riyadh. But I might in Baku, Astana and such oil-rich capitals of the former Soviet Union.

In a pessimistic view of the coming drama, the current behaviour patterns of nations might not offer encouragement that we could escape Henry Kissinger's 'contest', if it came to the sudden realisation that global oil production was collapsing, against all common expectations. If China and Japan cannot handle their current dispute over some probably valueless rocks in the East China Sea better than they do today, for example, what chance would nations have of managing conflict potential when it dawns on them that they are competing for rapidly diminishing supplies of a commodity that is the lifeblood of their economies?

Let me abandon the pessimistic view with that thought, and turn to the much more interesting prospect of an optimistic view.

In such a view, the exigencies of the oil crash will offer a unique window for world leaders to engage in urgent multilateral co-operation of the kind that successfully saw off the full imminent apocalypse potential of the 2008 financial crash. This time the opportunity would involve not bank bailouts but urgent collaborative clean energy deployment and energy-efficiency programmes: the acceleration of solution technologies as though mobilising for war.

How likely is it that they would do this? We have to bear in mind that world leaders, and their citizenry, would at this time be thinking in a different context. They would be able to take collective decisions within the 'power of context'.[3]

Let me give an example of what I mean from the 1930s. In that decade, Winston Churchill was warning the UK that the country should be preparing for war because a certain politician in Germany was up to no good. Few were prepared to listen to him at first. The context for their thinking was the party atmosphere of the times, and a pervasive dread of even mentioning war so soon after a conflagration that had killed millions. Churchill was offering a very uncomfortable narrative without any power of context. Once Hitler actually started invading neighbours, and his ruthless plan of conquest was clear, people had the power of context to aid their thinking. The choices became more obvious. And so my parents' generation began mobilising at speed. They found that they could churn out fighters, bombers, tanks and ships at a rate that took many of them by surprise.

This is the opportunity that world leaders will have, for a while, in the immediate aftermath of the oil crash caused by panic about peak oil. Clean-energy mobilisation is the task that bodies like the UK Industry Taskforce on Peak Oil and Energy Security argue for. This is why we want the government to work with industry to draw up a proper contingency plan. How much better could we do the task if ministries and industries work together on a national plan. How much better yet could governments do if they all work together on an international plan. I shall return to that point.

The 2010 Bundeswehr report on peak oil took a graphic line on the way people think about peak oil absent the power of context. It is 'difficult to imagine how significant the effects of being gradually deprived of one of our civilisation's most important energy sources will be', the German military analysts wrote. 'Psychological barriers cause indisputable facts to be blanked out and lead to an almost instinctive refusal to look into this difficult subject in detail. Peak oil, however, is unavoidable.'

It will be the trigger or triggers of panic about peak oil that will offer the power of context. The question will then become not whether society can or should mobilise a global energy economy that makes foreign oil dependency a thing of the past, but how fast can it do so.

***

The context for the question will be clear to all, after the oil crash: a massive rebuilding process, the re-engineering of national economies and a global system, to be achieved through a period of unprecedented societal stress, wherein the very cohesion of society will be in doubt.

Two very different world views will compete head-on in the answering of the question. The incumbency, with their 'new era of fossil fuels', will have suffered a setback as a result of the oil crisis, not a rout. They, like the investment bankers before them, will soon be arguing that the time for remorse is over. They will argue that all forms of national or regional carbon-fuel resource must now be mobilised as fast as possible. Their list will include accelerating some or all of the following, depending on individual national circumstances: remaining conventional oil and gas, including previously marginal fields, shale gas, tight oil, tar sands, gas-to-liquids, and of course coal, both for use in power plants so we can have more gas to turn to liquids, and to be turned itself into liquids using coal-to-liquids technology. The incumbency will say that it is regrettable about the climate consequences of this new fossil-fuel focus, but these consequences are even more subordinate to immediate needs now than they were before the oil crisis.

Many people and politicians will accept this, just as in 1939 many started immediately saying the British should let Hitler have his way in Europe, once he started invading neighbours.

The other world view will offer three arguments, probably in this order of importance in terms of appeal to politicians and publics. Together, within the power of context, they will make a far more compelling combination than they do today.

First, look at how *quickly* clean energy can be mobilised, especially energy efficiency.

Second, look at how much *less expensively* we can do this, over the lifetime of the new infrastructure we must now put in place.

Third, and only third – though this does remain the most important point for so many people, even given the awfulness of the oil crisis – look at the double-whammy for our money: how we dismantle climate risk as we make this mobilisation. After all, what really would be the point of mobilising a new era of fossil fuels if we end up subsequently destroying wealth as fast as we can create it via a climate meltdown?

Let me consider these three points in turn: the speed of the clean-energy mobilisation, the cost, and the double-whammy.

***

First, speed of mobilisation. Let me consider the historical mobilisation of renewable generation and renewable fuels by governments, the military, industry and communities; scaled experimentation using renewables at the national level; and modelling studies at the global, national and state levels. All these suggest that modern economies can be 100% powered by renewables, including in the transport sector, and far more easily, quickly and less expensively than many people think. Then I shall consider energy efficiency, which can be mobilised even faster and of course holds huge potential to shrink the amount of new renewable generation and renewable fuel use needed.

Among nations, German renewables provided more than 13% of energy and 29.9% of electricity in 2012. Renewables may provide only some 9% of the global energy mix, if we exclude traditional biomass, but they are growing fast even without the power of context that an oil crash would bring: 8.5% in 2012, a year in which almost 70% of new EU electricity generation was renewable, mostly wind and solar. Global investment in renewables, though it declined 12% from the record year of 2011, was still in excess of a quarter of a trillion dollars, and ahead of fossil fuels for the third successive year.[4]

In the optimistic security scenario, militaries could play as positive a role as they could a negative role in the pessimistic scenario. The US military has recently instigated a domestic programme for renewable fuels and power. On land, the army and air force are procuring renewables at scale.[5] At sea, the navy has an ambitious biofuels programme. This programme faces opposition in Congress, but US Navy Secretary Ray Nabus claims: 'The whole navy is committed to pursuing alternatives to foreign oil and the whole navy believes it is critical to our national security and combat capability. We simply have to figure out a way to get American-made, home-grown fuel that is stably priced, that is competitive with oil.'[6]

Such bullish sentiments are increasingly common in industry too, and across multiple sectors, some of which outperform governments in emissions reductions achievements and targets. As Part I shows, the retail sector can claim the best record to date. A major consideration in the business world is globalised trade, for which the future is questionable even in a high-oil-price world short of crisis. After an oil crash, oil prices would rise so high that transport costs would render much current trade uneconomic. Supply chains would need to be reconfigured in a hurry, to the extent possible in extremis. Companies with a high proportion of hydrocarbon-based products would face unprecedented supply problems.

Some who are currently planning for shortened supply chains may already be acting on the basis of such strategic considerations. IKEA is procuring solar roofs for all its stores worldwide – almost 300 of them in 26 countries – with the intention of being 100% renewably powered in all its global operations by 2020.[7] The company owns and operates 14 wind farms in six countries already. Others with active global programmes to shorten and decarbonise their supply chains include P&G and Unilever.

As for communities, Part I has shown examples of both the potential and the existing momentum of community renewables projects. In the rebuilding of economies and societies after the crash, local economic activity will need to be powered by locally sourced energy supply to a much greater extent than today. For those without significant domestic carbon-fuel reserves, most oil, gas and coal imports will become prohibitively expensive.

In Germany, in an economy pressured by over-reliance on Russian gas, with a voter-imposed nuclear ban, nearly 55% of the land area today comprises regions with declared energy-autonomy targets, intentions and processes in place. In this, the seeds of a localisation megatrend can already be detected.

Even for those with significant domestic carbon-fuel reserves, other options will be needed. For example, the Saudi Arabian government professes that it will solve its very own domestic, current, oil overconsumption crisis by rapid development of massive new home-grown industries in solar and nuclear.

The locally sourced energy of the post-crash world will probably need to be predominantly renewable, not nuclear. Supply chains can be built for renewables far faster in-country, or by pooling resources with near neighbours. In any event, nuclear is proving increasingly expensive, is being dropped as an option by key nations, and entails huge cooling-water challenges for countries – like Saudi Arabia – facing a water crisis as things stand. Finally, it is commonly thought that one more major accident will surely be the end for this industry globally, and in this context a recent review of the 12 main nuclear reactor meltdowns shows that on average one has happened every three years.[8]

In the last decade, the Germans have been expanding their renewable energy generation dramatically. A 2006 national real-time scaled experiment using elements of this renewable infrastructure, little known outside the country, showed that all German electricity could be provided, throughout the year, by solar photovoltaic and wind generation (78%) with some biogas generation (17%), and only a small amount of pumped-water storage of energy (5%): this without even factoring in hydropower, solar thermal, geothermal or marine power.[9] This experiment suggests that Germany would need 55 gigawatts of photovoltaics for an all-renewables supply; 25 gigawatts were already installed as of March 2012.

Encouraging as this discovery is, a trio of British, German and Italian scientists from solar research centres have recently shown that it would actually be easier to achieve in the UK the solar component of a 100% renewable supply.[10] This is because wind variation matches demand better. The UK has a peak electricity demand of about 60 gigawatts, with more than 12 gigawatts of renewables already on the grid. The all-renewable requirement for photovoltaics would be 37 gigawatts, 18 less than in Germany. Britain currently lags far behind Germany at little more than 1 gigawatt installed. But if the British installed photovoltaic capacity at rates Germany has already achieved, they could build enough to achieve solar's share of a 100% renewable supply by 2020. Britain, of course, also has significant largely untapped renewable assets in offshore wind, biogas, geothermal, hydropower and marine power. Based on these conclusions, the scientists have called for an immediate moratorium on all new non-renewable power plant construction in the UK. And this is a country with a Treasury planning generous subsidies for gas fracking, intending the nation to be turned into a gas 'hub', and calling for active suppression of renewables in the process.

With renewable electricity installation on such a scale, the spillover into non-carbon-based transport would surely not lag far behind. With the potential to charge vehicles at home or a place of work, automakers could reasonably expect sales of electric vehicles to explode. Cars themselves would become small mobile power plants, powered by batteries and fuel cells that would be charged by day in car parks and discharge into the grid by night. Growth of 'smart grid' technologies – use of computers to maximise the efficiency of electricity use and cut the need for supply – would be evolving in parallel. This scenario has the same decentralised modular attributes as the internet, and the same redundancy: knock out a part of it and the whole thing does not crash. This is why many people now speak of the 'energy internet'.

If all this is feasible in cloudy countries like Britain, or huge countries with much bigger hinterlands than coastal areas like Germany, then it might be easier still for sunny countries and countries with long coastlines.

Modelling studies have suggested that a modern global economy could be entirely powered by renewables in this way. As we have seen in Part I, the IPCC consensus view is that the global economy could be 80% renewable powered by 2050. Other studies are more bullish. Notably, a study by a team from Stanford University and the University of California at Berkeley demonstrates that wind, water and solar technologies could provide 100% of the world's energy as soon as 2030. This could be done, crucially, without mobilising any one technology any faster than technologies have already been mobilised historically.[11]

This needs to be emphasised: without mobilising any one technology any faster than technologies have already been mobilised historically.

Suppose we were actually to mobilise clean-energy technologies, and societal strategies, as fast as my parents' generation mobilised warplanes, warships and tanks in 1940? Suppose we were to do that under a regime of unprecedented multilateral co-operation outstripping the regime that governments put in place to bail out the banks?

How quickly could we then re-engineer the energy of nations?

That is an open question, but one that could be answered quickly by major governments serious about an emergency global clean-energy mobilisation in the face of the oil crisis, within the power of context. The answer would confound the current energy commentariat, I submit, thinking as they tend to do outside the power of context, and assuming as they so often do that history equals destiny when it comes to market penetration rates.

The energy world is rich in opportunities, along with the threats, for those prepared to lace their analyses of the future with the power of context. One obvious example is provided by solar PV manufacturing. At the time of writing, around half the world's 60 gigawatts of production capacity stands idle. This is a highly dysfunctional situation, especially given that even Saudi Arabia now regards solar PV as a strategically vital industry. It could yet worsen. The US and EU are both currently involved in trade disputes with China that could easily end in damaging outcomes for the industry. If we collectively mismanage crises like this, the question arises as to whether there will be enough survivors in the cleantech industries to provide a viable infrastructure for the renaissance. Stephan Dolezalek of Vantage Point Capital Partners puts it like this: 'Unless some portion of us survive the trip across the valley of death, there will be too little at scale for the transitioners to pick up on.' But at the time of writing there is every reason to be optimistic, he thinks. 'The last six months have seen the stock prices of companies like Tesla, First Solar, SolarCity and SunPower rise so much that among other things Total's investment in SunPower suddenly looks smart, unlike the disaster it seemed a year ago. With Tesla now at a $10 billion valuation, we are seeing the first Googles, Ciscos and Microsofts of the CleanTech era emerge.' To avoid trade wars capable of derailing promise such as this, how difficult would it be for the leaders of the major nations involved – the US, China and Germany – to sit down together and work out a treble-win rapprochement? Difficult perhaps today, in the light of grim realpolitik, though certainly not impossible. Very much easier after the oil crash, given the power of context.

What of the cost of re-engineering the global energy system to low- or zero-carbon? The first point to make is that cost is going to be viewed through different lenses after an oil crash. Value is going to enter play much more than it does today. But even where today's price structure is used as a basis for like-for-like comparison, renewable generation and fuels still win. Major authoritative studies that calculate it would be cheaper to go the low-carbon route than to maintain course with fossil fuels are numerous. Among those mentioned in Part I that do so are the 2006 Stern Review, the 2011 IPCC Special Report on Renewable Energy Sources and Climate Change Mitigation, and the Stanford/Berkeley study mentioned earlier in this chapter.

As for energy efficiency, and its ability to force-amplify renewables and renewable fuels, the guru of that discipline Amory Lovins and his Rocky Mountain Institute team were arguing compellingly that it was cheaper to save a barrel of oil than to produce it as long ago as 2004, the year the oil price began its extended climb.[12] In the 2012 *World Energy Outlook*, the IEA makes the crucial point that four-fifths of the energy efficiency potential in the buildings sector and more than half in industry still remain untapped, that these gains could be achieved without any major or unexpected technological breakthroughs, merely by taking actions that are economically viable and using investment that would be more than offset by reduced fuel expenditures, accruing huge GDP gains, especially in India, China, the United States and Europe.

Of course the incumbency tables its counter-narrative on a regular basis. For example, a coalition of gas firms infamously told the EU in February 2011 that the EU could meet its 2050 carbon targets more cheaply with gas than renewables. Gazprom, Centrica, Qatar Petroleum and others told the Commission it could save €900bn and still hit its 2050 carbon reduction targets if it built fewer wind farms and more gas plants. This just a few weeks after the Centrica CEO had predicted that gas prices would go nowhere but upwards in the long term, because the shale gas glut in America was unlikely to prove exportable.

The log on my website will give the reader looking for more detail in the ebb and flow of the 'We Are Cheaper / No You're Not' debate plenty to dig into. The main point I want to make here is that, after the crash, this manifestation of the energy civil war will look very different. Policymakers will have far more room to manoeuvre within the power of context. They will not be thinking about energy in the same way they do today.

That would be a good point to step into the next chapter, which is on the people and system implications of the oil crash.

But first I have to tidy that loose end and frequent afterthought that is climate change. I said there was a third argument that the zero- and low-carbon advocates would be deploying against the incumbency after the oil crash, involving a double-whammy. Let me make that case quickly, using military security arguments once again.

For years now I have watched the armed forces react with general concern to climate change, feeling growing frustration that they did not speak out more, and that when they do, governments don't seem to be listening. My experience began back in 1989. In October that year, shortly before the Berlin Wall fell, I held a seminar for an entire morning with senior British armed forces officers at a retreat of theirs. The major part of my brief was to talk to them about the then understanding of risk from global warming. I needn't have bothered about the risk. All they really wanted to do was talk about solutions to the problem. They needed no persuading that there was indeed a problem.

My hypothesis is that this was because there were ballistic missile submarine commanders in the room. They would have known even then that the Arctic ice cap was thinning worryingly. But of course, the data were all classified at the time.[13]

It took the British armed forces years to speak out forcibly in public. In June 2007, chief of the UK armed forces Air Chief Marshal Sir Jock Stirrup told a conference in London that climate change will increase competition for scarce resources and plunge many into desperate poverty, fanning conflict and terrorism across the world. 'It seems to me rather like pouring petrol onto a burning fire', he said.

In April 2008, the US Air Force took a similar view. William Anderson, an assistant secretary, called for an effort to combat climate change equivalent to the Apollo missions to put a man on the moon.

The terrible setback at Copenhagen happened the next year. The incumbency, it seemed, knew better even than the armed forces on both sides of the Atlantic.

Just as they say they do on oil depletion.

Chapter 20

The choice of roads

People and systems

How will we react, individually and collectively, to the crisis when it hits us? How will we manage the stresses in the aftermath? What will be the sum product of all those seven billion minds or, perhaps more materially, the very much smaller number of those minds empowered to make the key decisions?

After my first exposure to neuroscience and psychology in 2010, I have harboured an interest in the human brain or, at least, the aspect of it that is the mind: that complex of cognitive faculties that enable consciousness, perception, thinking, learning, reasoning and judgement. This will come as no surprise to the people who attended the seminar I describe in Part I. Among the things I have learned are these. The experimental evidence that we are 'predictably irrational' is now appreciable. It seems we are prone, among other hard-wired foibles, to an 'endowment effect': we overvalue what we already have and focus on this, rather than what we stand to gain via change. This trait applies to ideas as well as material things, which is why humans end up with ideologies that seem irrational in the face of clear real-world evidence. We tend to manifest great resistance to change, and feel compelled to keep options open. When we believe something will be good, our minds are capable of telling us it is good despite experience to the contrary. We have an optimism bias: we tend only to want to hear good news, and only tend to update on that basis. And so it goes on.[1]

The relevance of this to the five systemic risks I describe in this book will immediately be clear to the reader. Human evolution has wired the human brain slowly, over millennia, for cognitive decision-making suited to situations most unlike those we face today. For much of that time our predecessors had to deal with issues like encounters with other small groups of hunter-gatherers, quickly working out how to deal with them, or predators, and whether to run or gang up and fight them. After the rapid evolution of industrial society on the back of conventional energy use, the decision-making in modern human life is of a very different nature. And when things go wrong at the systemic level, our brains are not wired easily to accept the changes that, rationally, it is imperative we make.

There is one additional point that I think is germane to my history of the energy incumbency. Given the opportunity, psychologists know that many humans will tend to cheat. We tend readily to lie to others. We tend easily to lie to ourselves.

Again, let me not labour the point. I am an amateur in these matters, but I make it safe in the knowledge of the experimental basis for my thought.[2] I add this extrapolation of my own. Knowing all this about how the human mind works, it becomes easier to understand how the incumbency can persistently get away with outrageous excess, and their efforts to paint that excess as some kind of norm.

In view of this, and knowing the enormity of the denouements in the risks we are being collectively blind to – or choosing not to react effectively to – it is relatively easy to extrapolate the darkest possible outcome of our current multiple predicaments in energy. Literature and film help us.

In Cormac McCarthy's novel *The Road*, a man and his young son make a journey across a devastated post-apocalyptic America from the west to the east coast, through land almost bare of life. We are not told the reason for the apocalypse, but we can easily imagine the chain of events in the descriptions of the journey, and the man's tortured thoughts about his past.

Wildfires sweep the land periodically. We recall the terrible American droughts of recent years, the intense forest fire seasons, and how climate change, mostly fuelled by fossil-fuel burning, is predicted to escalate them. We learn that a huge explosion triggered the final collapse of utilities in the man's past domestic life. We think of the potential for war in our modern world, and how relatively easy it is to imagine hardline leaders in either China or America, or both, deciding on 'a contest' for the remaining accessible oil resources. We read of a poisoned landscape, lifeless rivers. We think of the reckless abandon with which the frackers pump toxic chemicals into the ground, and the tar sanders brew toxic lakes of waste water at such scale; how hard they fight to avoid any meaningful regulatory oversight; their default insistence that no harm can come to anyone or anything from their practices, these people who profess a deep-water blow-out impossible, even in the Gulf of Mexico, much less the Arctic. We recall how global warming devastates biodiversity, and how this would be compounded with other man-made assaults on nature, such as the terrible decline in bees, probably as a result of insecticide use, and the loss of their role as plant pollinators.

We read how the man and his son encounter rare desperate survivors, some in bands who have descended to unimaginable strategies in order to exist. We imagine how likely it would be that the psychopaths would survive more readily than others in such a lost society. We read of the emergence of cults, becoming progressively more extreme as society collapses, stage by stage. We recall how the human mind seems to allow the quick embracing of deeply ingrained belief systems, and then extremes of denial in defence of them. We remember a vast, excited crowd chanting 'Drill, Baby, Drill' at a political leader in a nation where every centre of scientific excellence is warning of ruinous climate change if all nations drill unconstrainedly. Where even its armed forces fear the environmental and social consequences of excessive drilling.

Such gloomy thinking makes unpleasant reading, but it will not be surprising to many anthropologists who have researched the rise and fall of past civilisations. Jared Diamond, who wrote the best-seller *Collapse: How Societies Choose to Fail or Succeed*, shows that powerful societies of the past thought that they were unique, right up to the moment of their collapse, and argues that we are essentially no different today. They could not adapt to, or even recognise, resource stresses. Neither can we, it seems, at least collectively or operationally. Reading Diamond's conclusions, and those of other experts in ancient civilisations like David Webster and Marcello Canuto, the patterns seem worryingly familiar. Even the timing is portentous. The Maya, Romans and Angkor of Cambodia lasted around 600 years. Western civilisation, according to some, began around 600 years ago.

Webster majors on lessons we should be learning from the history of the Maya and their rapid decline and demise some 800 years ago. 'In common with the Maya, we're not very rational in how we think about how the world works', he observes. 'They had their rituals and sacrifices. Magic, in other words. And we also believe in magic: that money and innovation can get us out of the inherent limits of our system, that the old rules don't apply to us.' When Canuto talks of the failure of their leaders – 'they did not respond correctly to a crisis which, in hindsight, was as clear as day' – I think of how today's presidents and prime ministers are going to look in the history books, once the dust settles on the great oil crash.[3]

Maybe, like earlier civilisations, we are collectively suicidal, and our risk blindness in the ways we use energy is just a manifestation of our sickness. There are surely others, like our use of fresh water, fish, forests.

Some people do accept this, if they are of a particular religious bent.

And yet it is not that difficult to evoke a counter-narrative, rooted in the real world, at the other end of the spectrum. Some have done so compellingly. Jeremy Rifkin weaves a wonderful vision of an increasingly empathic civilisation. He analyses the discoveries of neuroscientists about a more encouraging aspect of the human mind, what the practitioners I first heard in Oxford describe as the pro-social tendencies. Experimental studies show that the great majority of us tend to abhor violence and favour community. They show that groups who co-operate consistently do better than those wherein the individuals do not. They show that people tend to work harder when united by non-monetary social norms than they do for payment.

In Rifkin's analysis, every step change in energy use and means of mass communication has created greater opportunities for humans to exercise this tendency for empathy. In his view, modern social media serve to spread empathy outside historical in-groups at the same time that decentralised forms of renewable energy spread a greater propensity for community, and sharing, via the energy internet.

Let me paint on that same canvas from a personal perspective, beginning with the transformative power of a simple solar light, and extrapolating upwards in scale from there. In February 2013, BBC *Newsnight* filmed an interview with a young Kenyan, from a poor rural background, who had come 55th in the whole nation in his exams. He put his success down to the purchase of a single solar lantern, and the extra hours it gave him to do homework in the former darkness of his home. What does he want to do now? He wants to be a doctor.[4]

The allegory here needs no labouring. We think of this, and it is an easy next step to think of other elements of the history in Part I of this book: the buildings rapidly 100% powerable by solar and other renewables, the communities that are already *en route* to 100% renewable energy, the nations that think they can achieve the same, or nearly so. We think of the model-based studies that show how quickly the whole global economy could be run by renewables. We think of the transformative power such a new energy paradigm would hold: renaissance of community, improved human health, enhanced local and national security, and on and on – all made possible by the innate pro-social character of renewable energy, over and beyond its ability to cut greenhouse-gas emissions and eradicate oil dependency.

Then we imagine how world leaders, synergising and sharing in the emergency room of the global energy crisis, emboldened by the power of context – intent finally on individual and collective bids for true greatness – could work with these building blocks to break the mould of human history and its grubby endless repetition of past mistakes.

This is the kind of thing that the incumbency is struggling to stop happen. The discoveries of neuroscience suggest they are probably doing it at a collectively unconscious level. We can think of them as an institutionalised tribe of individuals with brains, like the rest of us, wherein only a bare minority of work even reaches the conscious level. Brains prone to the endowment effect, and all the rest of the predictably irrational human tendencies. Brains wired to cheat, if there is a chance of getting away with it.

As I listened to Tony Hayward speak at the 2012 *FT* summit, recalling all that I had seen of his history as described in this book, I reflected that I was entirely the wrong person to be on that panel with him. The perfect panelist would have been an eminent neuroscientist or psychologist, one conversant with the energy world. He or she could have given Tony, and the world, a neurological explanation for illusion-weaving in context.

***

Whatever the basis in neuroscience, one of the incumbency's most vital tools is manipulation of the media, and in this all the conventional energy industries have made and are making extensive use of black arts in their PR campaigning. Evidence for this is clear throughout the history in my book, and there is more in the log on my website. Let me emphasise the point with a little more of my own personal experience of it. In 2010, Shell invited me to write a short essay, which they said they would pay to have published in *Time* and *Fortune* magazines. I was appointed a CNN Principal Voice for that year, and this was part of the deal. I couldn't believe it. I suspected a rat. I wrote a tract railing against the oil and gas industry's serial irresponsibility on climate change and peak oil, their recklessness in exploiting the tar sands, their disinformation on oil reserves, and so on. I expected some effort to coerce me into editing it. There was none. The essay was published verbatim.[5]

I was bewildered. Later, an executive in a PR firm explained what was going on. He and his firm had better be nameless. He had worked on the Shell account, he told me over coffee, hoping that he could effect a little cultural change from within. He had failed and now was guilt stricken. He explained to me what the state of play was. His firm had conducted massive opinion surveying for Shell in multiple countries. The results couldn't have been clearer. The oil giant was hated by huge majorities right across the world. Not just disliked, hated. It would be a waste of time for them to try to rebuild any favourable image by advertising: they were so distrusted that nobody would believe a word they said. The same was true of other oil giants. If they grouped together to try to explain their case, the outcome would be even worse. People would feel ganged up on. So the PR agency recommended that the only option Shell had, if they cared at all about trying to rebuild their image, would be to associate their brand with the brands of discordant organisations: brands that questioned their world view, but were trusted by people.

And this was what they were doing. This was why they had the association with CNN. They were also engaged in multi-brand advertising with environment groups like the WWF.

This example is a window into both the level of sophistry that the incumbency is capable of and the ease with which it might be possible to begin disintermediating them, given the power of context. Without the power of context, society allows them to get away with the deployment of multiple millions of dollars pushing their toxic myth making. Yet we their customers hate them so much, evidently! It ought to be so easy to strip them of that right, in a rational world, to take their PR budgets away from them, and more: for example via windfall taxes, which we could then use to accelerate clean energy.

Come the oil crash, the power of context will create the space for politicians to take such survival-oriented measures, and with long-suppressed public backing.

Of course, the incumbency acts in ways that political players on the right tend to view favourably, for different reasons. In this way there is a parallel support process that is enormously powerful. The revelations in February 2013 about the creation of a vast climate denial network by conservative American donors make this point. Like so many incumbencies in history, the carbon-energy incumbency does not need to do all its dirty work alone.

The conservatives who fund climate denial do so because they see people who are concerned about mere climate as attackers on their 'endowment' in fossil fuels and industries that extract them. These people are resistant to change and want to conserve that endowment. Others even further to the right have a different agenda. They very much want change. In 2005, I addressed a public meeting on peak oil at which I was alarmed to find that the leadership of the fascist British National Party was in the audience.[6] During 2006, an undercover *Guardian* reporter discovered that the BNP planned to ride to power on the back of social and economic chaos arising from three issues: global warming, peak oil and the financial crisis.

It is easy to see why the fascists might lust after a resurgence of their fortunes in this way. It is not hard to imagine how economic impacts of either global warming or early peak oil could create armies of dispossessed and disenfranchised people for them to prey on. Interestingly, they also fund-raise with some success from the American far right.

Here we return depressingly to the kind of dismal landmark likely to have occurred at some point *en route* to Cormac McCarthy's dystopian vision in *The Road*. Yet it is a necessary thing to confront. Those unworried about the emergence of the surveillance society, or the steady infringements of human rights in countries like the UK, tend to emphasise the main potential upside: that we are making things more difficult for terrorists to perpetuate their dire acts. This would doubtless come as no surprise to neuroscientists, given what they know of the optimism bias. Others, however, worry that we are putting in place the perfect apparatus of repression in a police state. That apparatus – all the spying on private e-mails, the closed circuit TV cameras, the face-recognition software and computer databases, the police filming of people simply exercising their democratic right to protest – would be waiting for misappropriation, should the fascists come to power. We worry that the processes and even attitudes of policing are beginning to sit increasingly comfortably with this kind of future, as I have flagged in Part I. We worry that so many young French voted fascist in the 2012 French presidential election. We worry, as do the senior American armed forces officers cited earlier, that we are not doing anywhere near enough to address the root causes of terrorism, which of course have so much to do with our dependency on oil.

***

Here is how I think the power of context might come to our rescue as we endeavour to fashion a renaissance. As I write these words, I appreciate that there is a risk I harbour my own version of cognitive optimism bias. But I do think a good case can be made. It is rooted in five main premises, all interconnected and all feeding off each other in a way that makes the sum much greater than the parts. My five premises for the Road to Renaissance are as follows. First, the readiness of clean energy for explosive growth. Second, the intrinsic pro-social attributes of clean energy. Third, the increasing evidence of people power in the world. Fourth, the pro-social tendencies in the human mind. Fifth, the power of context that leaders will be operating in after the oil crash.

My contention is that there is plenty of evidence for each premise in the book thus far. I will consider each in turn, but first a caveat. I do not pretend that things won't get much worse before they get better. There will be rioting. There will be food kitchens. There will be blood. There already have been, after the financial crash of 2008. But the next time round will be much worse. In the chaos, we could lose our way like the Maya did. That is what many leading anthropologists expect to happen. The leadership issue will be particularly important in stopping it. We will need presidents and prime ministers keen to sit constructively in that multilateral emergency room, and we will need a critical mass of them – from nations big enough to make a difference – to co-operate as never before. We will need the masses to hold their leaders' feet to the fire. If we can manage that, the game of breaking the mould of human history will be on.

Premise One on the Road to Renaissance entails the readiness of clean energy for explosive growth. The next crash will lay bare all the incumbency's illusions about a new era of fossil fuels and of a wealth-creating financial system in need of only light-touch regulation. They will have left themselves at the mercy of a society that will be looking back in anger, and a political class that will feel impelled, given the state of their streets, to project the will of the people. Society will be being swept with a realisation that energy needs must be met in large measure at home, as fast as possible, and in a climate wherein modern financial institutions cannot in general be trusted with either individuals' money or the provision of financial services to viable economies.

The outcome will be much shaped by the extent to which people can believe, come the crash, that alternatives to the status quo are viable. Here the current state of play with clean energy will offer encouragement. For decades, the energy incumbency has – on the whole – been persuading society that alternatives to their wares are not for grown-ups. In recent years, as we have seen, the fast growth of many renewable technologies is proving otherwise for growing numbers of people.

In my day job, I encounter the 'seeing is believing' effect of hands-on experience of renewable energy ever more frequently. When neighbours see that a solar roof really works, even under cloudy skies, they tend to want one for themselves. A study at Stanford University has shown that Californian solar roofs are quite literally contagious, in this sociological sense. In pointing this out, I do not pretend that solar is a magic bullet. There are no magic bullets. It is just one member of the survival family, albeit an important one. But if it is being mobilised at full speed, along with the rest of the family, much will prove possible. Recall the McKinsey conclusion that simple projection of actual solar growth rates to date will, by 2020, change the face of the energy industry.

Energy-efficiency practitioners have even more encouraging stories to tell. One of my favourites involves a respected professor of energy studies in the UK, Catherine Mitchell. She hadn't fully understood the potential of energy efficiency, she claims, until she retrofitted her own home – a leaky ancient cottage. By fixing her windows and insulation, she saw her energy bills more or less disappear. In doing this, she came to realise that she was dealing not just with a challenge to the status-quo business model of the giant energy utilities, but also with a harbinger of their complete extinction.

Premise Two on the Road to Renaissance entails the intrinsic pro-social attributes of clean energy. They begin with the promotion of community interests. The inherent attributes of carbon fuels have done much to dismantle community, as any suburb and shopping mall shows. But in being more suited to local economic activity, clean energy favours the renaissance of community.

Within communities, local job creation will be vital as the economic damage caused by the oil crash is repaired. Clean-energy industries are significantly more job rich than the conventional energy industries, as the International Labour Organisation and the United Nations Environment Programme have shown: in solar's case, up to ten times more jobs per megawatt than gas.

Premise Three on the Road to Renaissance entails the increasing evidence of people power in the world. The Arab Spring leads the evidence for this premise, and gives a foretaste of what to expect after the crash. People will be looking back in anger in many theatres, and the exploding use of digital media will undoubtedly fan the forces of change. Of course, this will apply not just to the good, but also to the bad and the ugly. How that balance plays out will in turn depend on the other elements of the renaissance. The success of an organisation like Avaaz offers grounds for qualified optimism.

In the world of economics, the Occupy movement provides clear evidence of growing people power. As we saw in Part I, the experience of politicians in handling the 2011 Occupy Wall Street demonstrations is perhaps instructive of the things to come. After initially roundly condemning the movement, many politicians had to adjust their messages once opinion polls showed how many ordinary Americans shared the protesters' concerns, which the financial incumbency had sought to cast as extreme rabble-rousing.

We can expect people power to come into its own in the world of finance. Who would have thought that we would see a deputy governor of the Bank of England suggest that peer-to-peer lending and all the rest of the people-power innovations proliferating after the 2007 credit crunch and 2008 financial crash might disintermediate the main banks. But this, as we have seen, is exactly what Andy Haldane has said.

As for people power in the creation of autonomous resilient communities, there are over 2,000 Transition initiatives in 40 countries today. Many of them are already beginning to take concrete steps towards making their local economies more resilient, seeing community resilience as a vital form of economic resilience. They are even creating their own currencies, and in some instances these are being taken up quickly. The Bristol Pound, the complementary currency for the city of over 800,000 people, launched in September 2012 and is already accepted by hundreds of businesses in the city.

These may seem at first impression to be trivial examples, not likely to be much more than a sticking plaster on the dire injuries that society can expect to sustain in the wake of the oil crash. But when we sum this kind of evidence with trends in the corporate world, I suggest otherwise. For example, Google sought regulatory approval as an energy supplier as long ago as January 2010, and within less than a year was talking about investing in green energy as a 'mission-critical need'. Mission critical.

That is a good place to add the fourth and fifth premises on the Road to Renaissance. The fourth entails the pro-social tendencies in the human mind, and the fifth the power of context that leaders will be operating in after the oil crash. Add a grass-roots megatrend favouring the energy internet, add a corporate equivalent, brew them in a rising tide of empathic thinking, add the power of context, and transformative change can surely emerge, with the potential to proceed surprisingly fast. Just think how quickly Google emerged from a garage somewhere, in the course of the internet revolution.

Then imagine the leadership of Google and other giants of the corporate world, seeking to survive and prosper in the new world order, sitting with the American and Chinese leaderships not in an arena of trade bickering but one of urgent co-operation within a holistic international plan. Many things might be possible in that environment that today might be unimaginable, absent the power of context.

In an ideal scenario, nations will collaborate to such an extent in emergency mobilisation of clean energy that a whole new era of common security will emerge in the wake. In this narrative, growing numbers come to believe the view that their own security – whether at the national, communal or family level – is best guaranteed by guaranteeing the security of their neighbour. Strong leaders can do much to help that happen, and here I do not just talk of the leaders of nations, but the leaders of companies, institutions, communities and citizen organisations.

If we get the renaissance right in this way, then common security thinking spreads with the speed of a viral infection, and today's vast military budgets begin flowing into social themes. If renewables are mobilised as fast as I suggest is possible, then the current vast subsidies for conventional energy can be redeployed in more socially constructive ways.

Could it happen? Much will depend on this tendency to empathy in the human mind, both in the minds of leaders and in the minds of the ordinary citizens, and the pressures they put on their leaders in the days that will decide whether we take the Renaissance Road or the Maya Road.

***

I shall leave the Maya Road to Hollywood from this point on. The Renaissance Road interests me far more.

If we do take that one, where might it all end up?

In the aftermath of the 2008 financial crash, many commentators came to conclude that far-reaching reforms were necessary in the capital markets. The fact that the *Financial Times* could run a series entitled 'Capitalism in Crisis' speaks volumes. Part I features others, and there are many more in the log on my website. Let me extract two to tee up my next point. Who would have thought, pre-crash, that Jack Welch, legendary ex-CEO of GE, would say in 2009 that focusing on quarterly profit is 'a dumb idea' for executives? In saying that, he essentially turned his back on everything he did to become a legend. Or that the serving CEO of huge retail group Kingfisher, Ian Cheshire, could tell the world in 2011 that 'we need a radical reappraisal of capitalism' without being fired by investors?

My point is this. If this is the direction of the zeitgeist before the oil crash, just imagine what might be possible, given an explosion of people power, within the power of context, afterwards.

Governments would direct both their own procurement and their reforms of conventional capital at massive mobilisation of social infrastructure projects – of which energy would be just a part – creating millions of new jobs in the process. The Green New Deal group has tended to analyse and advocate this idea in a national context. Suppose it were to be internationally co-ordinated?

In the wider business world, brands would find themselves competing for the trust of the new consumer. Pre-crash surveys already show responsible businesses outperforming their rivals in both good and hard times. It is far from a stretch to imagine many companies post-crash restructuring for the well-being of all their stakeholders, not for just a few short-termist share-holders and their brokers.

I think of such companies as renaissance companies. In our vocational lives, I and my close colleagues are trying to create a microcosm of what to expect post-crash. We are currently designing a new company, SunnyMoney – currently wholly owned by our charity SolarAid – as a hopeful standard bearer for this new class of company. SunnyMoney has become the lead retailer of solar lanterns in Africa in just a few years, and our soaring sales offer us the chance to be a Coca-Cola of the microsolar world. But what is the point, we believe, if we can ultimately be sold for the enrichment of a few investment bankers and venture capitalists to some dreary electronics giant or the like? We have not gone to such people for capital, but rather to a wholly new class of investor: the type of people plugged into the innate sense of empathy in their minds. We hope to raise much of the working capital we need via crowdfunded debt, and are already borrowing from ordinary people at 5% interest. This is a rate higher than any major bank offers on its so-called savings accounts, but it is much lower than the usurous rates at which financial institutions lend to ordinary citizens.

What other interventions could help achieve the kind of snowballing change I am describing on the Road to Renaissance? One area is phil-anthropy. The total endowment of the world's philanthropic foundations amounts to a staggering $800 billion dollars, or thereabouts. Historically, these foundations have invested their shares of this sum in traditional places, and given the interest out as grants aiming to foster social change. But with the climate crisis being as bad as it has become, a question arises. What is the point of sitting on this vast sum in a world irredeemably on course for a six-degree temperature rise and beyond, or taking the Maya Road after the next crash? There will be little that can be done with endowments in that meltdown. Why not, therefore, table it in one go, today, to pump into carefully chosen organisations constructing the Road to Renaissance? How would foundation leaders answer if their children were to ask in 2025, say, in a world staring down the barrel of a runaway greenhouse effect, why they had not thrown their hundreds of billions into the fight when there was still a chance?

If the financial system does reconfigure the way the more bullish crowdfunders expect, the foundations could reasonably expect to earn a large part of their endowments back from the citizenry of the renaissance, in returns on investment.

In a climate where society was re-engineering itself in the ways I describe, on multiple fronts, in multiple countries, governments would very soon find themselves under pressure to rethink the very basis of capitalism. They would find themselves in negotiation – in a kind of post-crisis Bretton Woods – for core reform of economics itself.

Today, prosperity is understood as a successful, flourishing or thriving condition: simply, a state in which things are going well for us. Every day the system in which we live tries to persuade us – via TV news, politicians' speeches, corporate pronouncements, inducements to consume and so on – that our prosperity is intimately linked to whether or not gross national product is growing and whether stock markets are riding high. These are the two main measuring sticks for the version of capitalism on which most countries base their economies as the crash looms.

Other ways of measuring prosperity, such as employment and savings, follow these two. If GNP – the total national output of goods and services – is in recession, then unemployment will rise, and that means growing numbers of unprosperous people without salaries. If stock markets are falling, that means falling pension values, and rising numbers of unprosperous people in retirement.

So questioning the primacy of growth-at-all-costs to date has been deemed to be the act of lunatics, idealists and revolutionaries. The energy incumbency stands in the front ranks of those who cast the stones. But question it we must, and question it we would, on the Road to Renaissance.

In fact we have started. A 2008 book by Tim Jackson, formerly economics commissioner in the UK government's now-disbanded Commission for Sustainable Development, perhaps marks a beginning. He called his book *Prosperity without Growth*. In the wake of the 2008 financial crisis, some surprising people came to echo his title. President Sarkozy, the Nobel-prizewinning economist Joseph Stiglitz and elements of the *Financial Times*'s commentariat were among those who began arguing that prosperity is possible without GNP growth, and indeed that prosperity will soon become impossible *because of* GNP growth.

Jackson and his colleagues on the Sustainable Development Commission made the relevant economic arguments understandable to the lay reader. In essence, they argued, the idea of a non-growing economy may be anathema to an economist, but the idea of a continually growing economy is anathema to an ecologist.

This is the core of the debate. Endless growth is a ridiculous notion to the typical ecologist because we live on a planet with finite resources, the mining and use of some of which are undermining our planet's life-support systems. But the typical economist believes we can 'decouple' GNP growth from resource use through an increased efficiency intrinsic to capitalism: that we can grow our economies and reverse environmental degradation too.

My hope for *The Energy of Nations* is that I have helped expose that thinking as a dangerous myth: the kind of 'magic' the Maya believed in.

As a creature of modern capitalism, a venture-capital-backed energy-industry boss, a private equity investor, and all the rest of it – a former believer in the magic – I am now convinced that capitalism as we know it is torpedoing our prosperity, killing our economies and threatening our children with an unliveable world. It needs to be re-engineered, root and branch.

And, I believe, that is how far the Road to Renaissance would necessarily have to go. A very long way from where we are today. A place rendered both desirable and feasible by the power of post-crash context. If it does, then modern capitalism's worst ever crash may prove to be a cloud drifting across human history that has a very big silver lining indeed.

Notes

Prologue

1 'Wall Street e-mail trail overview', PBS. http://www.pbs.org/now/politics/wallstreet.html
2 When I talk of the oil price in this book, I refer to the benchmark known as Brent Crude. There are other benchmarks, wherein prices differ, usually only slightly, the most commonly used being West Texas Intermediate.

I Lies, scaremongering and affordable oil

1 'Shell admits it misled investors', *Guardian*, 20 April 2004. http://www.guardian.co.uk/business/2004/apr/20/oilandpetrol.news1
2 The Justice Department and the FSA both decided later not to proceed (July and November respectively).
3 'Coping with sky high oil prices', *Business Week*, 30 August 2004. http://www.businessweek.com/stories/2004-08-29/coping-with-sky-high-oil-prices
4 'Once seen as an alarmist fear, an attack on key Saudi oil terminal could destabilise West', *Guardian*, 3 June 2004. http://www.guardian.co.uk/world/2004/jun/03/saudiarabia.oil
5 'Oil threat to world economy', *Guardian*, 5 August 2004. http://www.guardian.co.uk/business/2004/aug/05/oilandpetrol.politics
6 'BP chief's claims at odds with rivals', *Financial Times*, 17 September 2004. No url.
7 'Pouring oil on troubled economists', *Observer*, 10 October 2004. http://www.guardian.co.uk/business/2004/oct/10/politics.oilandpetrol
8 'Browne calms oil supply fears', *Guardian*, 27 October 2004. http://www.guardian.co.uk/business/2004/oct/27/oilandpetrol.news

2 Under the volcano

1 Matthew Simmons, *Twilight in the Desert: The Coming Saudi Oil Shock and the World Economy* (Wiley, 2005).
2 http://www.eia.gov/countries/country-data.cfm?fips=SA
3 Jeffrey Robinson, *Yamani: The Inside Story*, Simon & Schuster, 1988, p. 102.
4 For an account of Harlan Watson at work in the climate negotiations, and my interactions with him, see Jeremy Leggett, *The Carbon War* (Penguin, 2000), pp. 128–31.
5 *The Carbon War* had to be read carefully by the UK's leading firm of libel lawyers before Penguin would publish it.

3 Doomed to failure

1 'Stern Review on the economics of climate change', HM Treasury, 2006. http://webarchive.nationalarchives.gov.uk/+/http:/www.hm-treasury.gov.uk/sternreview_index.htm
2 'Climate change 2001: the scientific basis', Working Group 1 Contribution to the Intergovernmental Panel on Climate Change Third Assessment Report, 2001. http://www.grida.no/publications/other/ipcc_tar/
3 'Plenty of oil? Just drill deeper', *Business Week*, 17 September 2006. http://www.businessweek.com/stories/2006-09-17/commentary-plenty-of-oil-just-drill-deeper
4 International Energy Agency, *World Energy Outlook 2006*, November 2006. http://www.iea.org/publications/freepublications/publication/name,3650,en.html
5 The url for this article is no longer on the *FT*'s website, and the quotes come from the Triple Crunch Log on my website www.jeremyleggett.net. But see an *FT* editorial from 8 November 2006, which discusses the 'doomed to failure' quote. http://www.ft.com/cms/s/0/2f9892ca-6ed0-11db-b5c4-0000779e2340.html#axzz2P7O6bZFy
6 Jeremy Leggett and Ian Vann, 'Are we running out of oil?', Environment on the Edge debate, Cambridge University, 30 November 2006. UNEP Our Planet transcript: www.ourplanet.com/imgversn/edge/. . ./Leggett%20and%20Vann.pdf
7 Not to be confused with tight oil – oil fracked from shale – which has yet to feature in the history.

4 We are not responsible

1 'Iran actually is short of oil', *International Herald Tribune*, 8 January 2007. http://www.nytimes.com/2007/01/08/opinion/08iht-edstern.4136795.html?_r=0

2 'Climate change 2007: the physical science basis', Working Group 1 Contribution to the Intergovernmental Panel on Climate Change Fourth Assessment Report, Technical Summary, 2 February 2007. http://www.ipcc. ch/publications_and_data/ar4/wg1/en/contents.html

3 This section was published on the *Guardian*'s website and has been only slightly edited here. Jeremy Leggett, 'No time at all', *Guardian*, 2 May 2008. http://www.guardian.co.uk/commentisfree/2008/may/02/washingtheirhands

5 The risk of contingency

1 'IEA: without Iraqi oil, we'll be in deep trouble by 2015', *Le Monde*, 28 June 2007. Translation posted on the *Oil Drum*, Europe: http://europe.theoildrum. com/node/2721#more

2 'US urged to act on energy', Ed Crooks, *Financial Times*, 17 July 2007.

3 For further explanation see Euan Mearns, 'Peak oil – now or later? A response to Daniel Yergin', *Oil Drum*, September 2011. http://www.theoildrum.com/ node/8391; Mark Lewis and Michael Hsueh, 'Crude Oil: Iceberg Glimpsed Off West Africa', Deutsche Bank, Global Markets Research, Special Report, 2 February 2012.

4 'Citigroup chief stays bullish on buyouts', *Financial Times*, 9 July 2007. http:// www.ft.com/cms/s/0/80e2987a-2e50-11dc-821c-0000779fd2ac.html#axzz2 ILOuy8kc

6 The small print

1 'Big Oil has trouble finding new fields', *San Francisco Chronicle*, 1 February 2008. www.sfgate.com/business/article/Big-Oil-has-trouble-finding-new-fields- 3230034.php

2 'Action needed to avoid oil crisis', *Oil and Gas Journal*, 18 February 2008. No url.

3 Fatih Birol, 'Outside view: we can't cling to crude; we should leave oil before it leaves us', *Sunday Independent*, 2 March 2008.

4 'World faces "oil crisis"', *Bloomberg*, 11 June 2008. http://www.bloomberg. com/apps/news?pid=newsarchive&sid=a0SE24WXEk5U

5 This section was published on the *Guardian*'s website and has been only slightly edited here. Jeremy Leggett, 'Spoiling the barrel', *Guardian*, 13 June 2008.

7 When the dancing stops

1 'Confessions of a sub-prime mortgage baron', *Guardian*, 19 September 2008. http://www.guardian.co.uk/business/2008/sep/19/subprimecrisis.richardbitner

2 'When the junk was gold', *Financial Times* magazine, 17 October 2008. http:// www.ft.com/cms/s/0/65892340-9b1a-11dd-a653-000077b07658.html#axzz 2KyD4bNw7

3 'Hedge fund manager slams bankers', *Financial Times*, 17 October 2008. http://www.ft.com/cms/s/0/b0a40c72-9c83-11dd-a42e-000077b07658.html #axzz2KyD4bNw7

4 'Saudi King says keeping some oil finds for the future', *Reuters*, 13 April 2008. http://uk.reuters.com/article/2008/04/13/saudi-oil-idUKL139687720 080413

5 UK Industry Taskforce on Peak Oil and Energy Security, 'The oil crunch', 29 October 2008. http://peakoiltaskforce.net/download-the-report/download-the-2008-peak-oil-report/

8 This house believes

1 'Bankers kept silent over Madoff fears', James Doran, *Observer*, 15 February 2009. http://www.guardian.co.uk/business/2009/feb/15/madoff-fraud-wall-street

2 'Banking's big question: why didn't anyone stop them?' Nick Mathiason, Heather Connon and Richard Wachman, *Observer*, 15 February 2009. http://www.guardian.co.uk/business/2009/feb/15/banking-recession

3 'Bank faces probe over "threats" to directors', Toby Helm, Jamie Doward and Paul Kelbie, *Observer*, 22 March 2009. http://www.guardian.co.uk/business/2009/mar/22/rbs-threats-directors-lord-foulkes

4 Julian Rush summarises this debate in the October 2009 issue of *Petroleum Review* magazine, in which feature both the David Jenkins and Jeremy Leggett statements. http://www.jeremyleggett.net/2009/11/geologists-vote-that-peak-oil-is-a-concern/

9 We will be blamed forever

1 'At the core of this policing crisis is a leadership failure', David Gilbertson, *Guardian*, 20 April 2009. http://www.guardian.co.uk/commentisfree/2009/apr/20/policing-relations-general-public

2 'Sun sets on BP's hopes', Ed Crooks, *Financial Times*, 13 May 2009. http://www.ft.com/cms/s/0/d7b2a18e-3ff3-11de-9ced-00144feabdc0.html#axzz 2P7O6bZFy

3 An account of this event can be read in 'Future of energy debate – Groningen', Digital Energy Journal, 1 September 2009. http://www.digitalenergyjournal.com/n/Future_of_energy_debate_Groningen/6351fc55.aspx

4 'Financial Services Authority chairman backs tax on "socially useless" banks', *Guardian*, 27 August 2009. http://www.guardian.co.uk/business/2009/aug/27/fsa-bonus-city-banks-tax

5 'Bank of England says financiers are fuelling an economic doom loop', *Daily Telegraph*, 6 November 2009. http://www.telegraph.co.uk/finance/financetopics/financialcrisis/6516579/Bank-of-England-says-financiers-are-fuelling-an-economic-doom-loop.html

6 'Energy security: a national challenge in a changing world', Malcolm Wicks MP, published by the Department of Energy and Climate Change, August 2009. No url.

7 The Taskforce chairman, Will Whitehorn of Virgin, and I wrote an op-ed in the *FT* about this: 'Do not discount the threat of peak oil', *Financial Times*, 9 August 2009. http://www.ft.com/cms/s/0/bed9186c-8514-11de-9a64-001 44feabdc0.html#axzz2L3G2p1VO

8 'Total CEO expects higher crude prices, supply squeeze in 2014', *Bloomberg*, 11 September 2009. http://www.bloomberg.com/apps/news?pid=email_en& sid=aTv00Tc4yhSA

9 'Key oil figures were distorted by US pressure, says whistleblower', *Guardian*, 9 November 2009. http://www.guardian.co.uk/environment/2009/nov/09/ peak-oil-international-energy-agency

10 This account is based on blog posts that appeared over the period of time described in the *Financial Times*, in particular: http://blogs.ft.com/energy-source/2009/12/17/climate-experts-forum-the-leaders-arrive-is-a-copenhagen-climate-change-deal-closer/ http://blogs.ft.com/energy-source/2009/12/18/ climate-experts-forum-who-is-responsible-for-the-chaos/ http://blogs.ft.com/ energy-source/2009/12/19/climate-experts-forum-the-copenhagen-agreement-a-disappointment-or-a-relief/

10 As bad as the credit crunch

1 'Diamond lashes out at Obama bank plans', *Financial Times*, 27 January 2010. http://www.ft.com/cms/s/0/f58ce0bc-0b30-11df-9109-00144feabdc0.html #axzz2L3G2p1VO

2 'The oil crunch: a wake-up call for the UK economy', Second Report of the UK Industry Taskforce on Peak Oil and Energy Security, 8 February 2010. http://peakoiltaskforce.net/download-the-report/2010-peak-oil-report/

3 http://www.eia.gov/forecasts/steo/report/global_oil.cfm

4 Richard Branson address at the press conference launching the UK Industry Taskforce on Peak Oil and Energy Security 2010 report, Royal Society, London, 8 February 2010. http://www.youtube.com/watch?v=16Qo8tvL3YI

5 Jeremy Leggett address at the press conference launching the UK Industry Taskforce on Peak Oil and Energy Security 2010 report, Royal Society, London, 8 February 2010. http://www.youtube.com/watch?v=kaxJum8FD5g

6 Chris Barton address at the press conference launching the UK Industry Taskforce on Peak Oil and Energy Security 2010 report, Royal Society, London, 8 February 2010. http://www.youtube.com/watch?v=Dyj_FZOtNTg

7 'Energy minister will hold summit to calm rising fears over peak oil', *Observer*, 21 March 2010. http://www.guardian.co.uk/business/2010/mar/21/peak-oil-summit

8 'Joint Operating Environment 2010', US Joint Forces Command, March 2011. http://www.jfcom.mil/newslink/storyarchive/2010/pa031510.html; 'US military warns oil output may dip causing massive shortages by 2015', *Guardian*, 11 April 2010. http://www.guardian.co.uk/business/2010/apr/11/peak-oil-production-supply

9 'Fears raised over process of extraction', *Financial Times*, 7 March 2010. http://www.ft.com/cms/s/0/ccd8412a-2a11-11df-b940-00144feabdc0.html #axzz2L9Oc8V6l

10 'The neuroscience of change: understanding the brain, influencing behaviour', Expert seminar and discussion at the Skoll World Forum on Social Entrepreneurship, Oxford, 15 April 2010. Audio: http://skollworldforum.org/session/the-neuroscience-of-change-understanding-the-brain-influencing-behaviour/

11 'Chevron chief shuns shale gas rush', *Financial Times*, 25 April 2010. http://www.ft.com/cms/s/0/272d6046-5088-11df-bc86-00144feab49a.html# axzz2L9Oc8V6l

12 'Are policymakers, economists and peak oilists starting to speak the same language?' *Financial Times*, 21 April 2010. http://blogs.ft.com/energy-source/2010/04/21/are-policymakers-economists-and-peak-oilists-starting-to-speak-the-same-language/

11 You are the flip side of austerity

1 'BP boss admits job on the line over Gulf oil spill', *Guardian*, 14 May 2010. http://www.guardian.co.uk/business/2010/may/13/bp-boss-admit's-mistakes-gulf-oil-spill

2 'BP CEO: "I'd like my life back"', *The Hill*, 31 May 2010. http://thehill.com/blogs/e2-wire/e2-wire/100735-bp-ceo-id-like-my-life-back

3 'Goldman Sachs settles with SEC', *Financial Times*, 16 July 2010. http://www.ft.com/cms/s/0/4bd43894-904c-11df-ad26-00144feab49a.html# axzz2L9Oc8V6l

4 'Why it is still bank business as usual', *Financial Times*, 10 September 2010. http://www.ft.com/cms/s/0/15c88654-bd0d-11df-954b-00144feab49a.html #axzz2LQiqcROy

5 'Barclays' Bob Diamond hits out at criticism of "casino banks"', *Guardian*, 12 September 2010. http://www.guardian.co.uk/business/2010/sep/12/barclays-bob-diamond-casino-banks

6 'Former FSA boss Hector Sants joins Barclays as head of compliance', *Guardian*, 12 December 2012. http://www.guardian.co.uk/business/2012/dec/12/former-fsa-hector-sants-barclays-compliance

7 'Rohstoffknappheit: Bundeswehr-Studie warnt vor dramatischer Ölkrise', *Spiegel*, 31 August 2010. http://www.spiegel.de/wirtschaft/soziales/0,1518,714878,00.html

8   http://translate.googleusercontent.com/translate_c?hl=fr&ie=UTF-
    8&sl=de&tl=en&u=http://www.spiegel.de/wirtschaft/soziales/0,1518,71487
    8,00.html%23ref%3Drss&rurl=translate.google.com&twu=1&usg=ALkJrhg
    M6akZvTmnkf-cFkgH7hm_YMd0tQ
9   'Peak oil alarm revealed by secret official talks', Observer, 22 August 2010.
    http://www.guardian.co.uk/business/2010/aug/22/peak-oil-department-
    energy-climate-change
10 http://www.greennewdealgroup.org/
11 'Coal India IPO shows the mountain we have to climb', Jeremy Leggett,
    Guardian, 9 November 2010. http://www.guardian.co.uk/sustainable-business/
    blog/coal-india-ipo-climate-change

12 Houston, it's just possible we may have a problem

1   'Saudi to develop solar and nuclear power', Financial Times, 24 January 2011.
    http://www.ft.com/cms/s/0/1e31d396-27a6-11e0-a327-00144feab49a.html#
    axzz2LWRtSZ4n
2   http://www.publications.parliament.uk/pa/cm201011/cmhansrd/cm110
    127/text/110127w0002.htm
3   'UK industry must press on, says EDF chief', Financial Times, 17 March 2011.
    http://www.ft.com/cms/s/0/6e1ba3f2-50df-11e0-9227-00144feab49a.html#
    axzz1GwRf0qD2
4   'Germany debates how to dump nuclear power', Reuters, 15 April 2011.
    http://www.reuters.com/article/2011/04/15/us-germany-nuclear-idUSTRE
    73E2I920110415
5   'WikiLeaks cables: Saudi Arabia cannot pump enough oil to keep a lid on
    prices', Guardian, 8 February 2011. http://www.guardian.co.uk/business/
    2011/feb/08/saudi-oil-reserves-overstated-wikileaks
6   'U.S. must help Saudi Arabia turn into "the Saudi Arabia of Solar", WikiLeaks
    cables say', Jeremy Leggett, Huffington Post, 14 February 2011. http://www.
    huffingtonpost.com/jeremy-leggett/us-must-help-saudi-arabia_b_822787.
    html
7   'Oil crunch: are we there yet?', ABC TV, Catalyst, 28 April 2011. The programme
    can be viewed online at: http://www.abc.net.au/catalyst/stories/3201781.htm

13 The anti-Oil Shock Response Plan plan

1   Ian Cheshire, 'Imagining a new, sustainable capitalism', Guardian, 24 March
    2011. http://www.guardian.co.uk/sustainable-business/blog/kingfisher-ceo-ian-
    cheshire-sustainable-capitalism
2   Matt Taibii, 'The people vs. Goldman Sachs', Rolling Stone, 11 May 2011.
    http://www.rollingstone.com/politics/news/the-people-vs-goldman-sachs-
    20110511

3 'Eon takes up Merkel move to go green', *Financial Times*, 26 June 2011. http://www.ft.com/cms/s/0/473c525c-a014-11e0-a115-00144feabdc0. html#axzz1QPS5kIcA

4 'S'accrocher au nucléaire, une erreur stratégique', *Libération*, 13 June 2011. http://www.liberation.fr/terre/01012343176-s-accrocher-au-nucleaire-une-erreur-strategique

5 'Worst drought in more than a century strikes Texas oil boom', *Bloomberg*, 13 June 2011. http://www.bloomberg.com/news/2011-06-13/worst-drought-in-more-than-a-century-threatens-texas-oil-natural-gas-boom.html

6 'Insiders sound alarm amid a natural gas rush', *New York Times*, 25 June 2011. http://www.nytimes.com/2011/06/26/us/26gas.html?_r=1

7 'Britain's first solar powered town, episode three', YouTube film published on 5 September 2011. http://www.youtube.com/watch?v=0kjA-wR8zxo

8 'Britain's first solar powered town, episode four', YouTube film published on 2 March 2012. http://www.youtube.com/watch?v=GEklAJ1sc2o&feature=endscreen&NR=1

14 A 'bollocks' subject

1 'Greenhouse-gas emission targets for limiting global warming to 2°C', Malte Meinshausen, Nicolai Meinshausen, William Hare, Sarah C.B. Raper, Katja Frieler, Reto Knutti, David J. Frame and Myles R. Allen, *Nature*, March 2009. http://www.nature.com/nature/journal/v458/n7242/full/nature08017.html

2 Euan Mearns, 'Peak oil – now or later? A response to Daniel Yergin', *Oil Drum*, 22 September 2011. http://www.theoildrum.com/node/8391

3 'BP's bid to clean up its act dealt blow by revelations in Russia case', *Guardian*, 5 November 2011. http://www.guardian.co.uk/business/2011/nov/05/bp-clean-up-russia-case

4 Charles Hendry MP and Jeremy Leggett debate, YouTube, 5 November 2011, uploaded 6 November 2011. http://www.youtube.com/watch?v=My-nKK su2pQ

5 'Harvard students, citing economic inequality, stage walkout', *Bloomberg Businessweek*, 8 November 2011. http://www.businessweek.com/bschools/blogs/mba_admissions/archives/2011/11/harvard_students_citing_economic _inequality_stage_walkout.html

15 To the point of being suicidal

1 'Oil will decline shortly after 2015, says former oil expert of International Energy Agency', *Oil Drum*, 5 January 2012. Interview by Matthieu Auzanneau. This article previously appeared in *Le Monde*. http://www.theoildrum.com/node/8797

2 'Fracking boom could finally cap myth of peak oil', *Bloomberg*, 1 February 2012. http://www.bloomberg.com/news/2012-02-01/fracking-boom-could-finally-cap-myth-of-peak-oil-peter-orszag.html

3 Mark Lewis and Michael Hsueh, 'Crude oil: iceberg glimpsed off West Africa', Deutsche Bank, Global Markets Research, Special Report, 2 February 2012.

4 'Modern capitalism: an energy entrepreneur's appraisal after twelve years in the markets', presentation to the Breakthrough Capitalism conference, London, 29 May 2012, posted on YouTube 10 June 2012. http://www.you tube.com/watch?v=77jmQ762hzo

16 A new era of fossil fuels

1 'Familiar echoes in shale gas boom', *Financial Times*, 6 May 2012. http://www.ft.com/cms/s/0/75942e5c-944e-11e1-bb0d-00144feab49a.html#ixzz1uGJ6G3C6

2 'Learning from US shale gas experiences', presentation by Art Berman to the annual conference of the Association for the Study of Peak Oil, Vienna, 30 May 2012. http://www.aspo2012.at/conference-presentations/day1part2/

3 For the reader seeking more detail, I summarise papers on this aspect of the shale gas and tight oil boom, including papers by analysts such as Chris Nelder and Rune Likvern, on my website.

4 'Exxon: "losing our shirts" on natural gas', *Wall Street Journal*, 27 June 2012. http://online.wsj.com/article/SB100014240527023035615045774925010266260464.html

5 'Shaming the banks into better ways', *Financial Times*, 28 June 2012. http://www.ft.com/cms/s/0/6dc5b9a2-c117-11e1-853f-00144feabdc0.html#axzz2NEN4L4fJ

6 Bob Diamond, *Today* Business Lecture, undated 2012. http://news.bbc.co.uk/today/hi/today/newsid_9630000/9630673.stm

7 'Bare-faced bankers should be treated as criminals: prosecuted and imprisoned', *Guardian*, 20 July 2012. http://www.guardian.co.uk/commentisfree/2012/jul/20/bare-faced-bankers-criminals-prosecuted

8 'The weekly wrap', Steve Levine, *Foreign Policy*, 27 July 2012. http://oiland glory.foreignpolicy.com/posts/2012/07/27/the_weekly_wrap_july_27_2012_part_i?wp_login_redirect=0

9 'Nuclear "hard to justify", says GE chief', *Financial Times*, 30 July 2012. http://www.ft.com/cms/s/0/60189878-d982-11e1-8529-00144feab49a.html#axzz2NRp42ssP

10 Note that different forms of generation have different capacity factors: ratios of actual output, over a period of time, to potential output if it were possible to operate at full nameplate capacity indefinitely.

11 'Are we entering a new fossil fuel era?' Video of discussion panel at the *FT* Global Energy Leaders Summit, 18 September 2012. https://www.ft-live.

com/ft-events/ft-global-energy-leaders-summit/sessions/panel-are-we-entering-a-new-fossil-fuel-era

17 More unhinged by the week

1 Raymond Pierrehumbert, 'The myth of "Saudi America": straight talk from geologists about our new era of oil abundance', *Slate*, 6 February 2013. http://www.slate.com/articles/health_and_science/science/2013/02/u_s_shale_oil_are_we_headed_to_a_new_era_of_oil_abundance.html

2 Euan Mearns and Rembrandt Koppelaar, 'Oil watch: drill, baby, drill', *Oil Drum*, 30 January 2013. http://www.theoildrum.com/node/9795

3 David Hughes, 'Drill, baby, drill: can unconventional fuels usher in a new era of energy abundance?', Post Carbon Institute, 18 February 2013. http://www.postcarbon.org/drill-baby-drill/

4 Deborah Rogers, 'Shale and Wall Street: was the decline in gas prices orchestrated?', Energy Policy Forum report, 19 February 2013. http://energypolicyforum.org/portfolio/was-the-decline-in-natural-gas-prices-orchestrated/

5 'Banking industry's year of shame ends in a blizzard of Libor revelations', *Observer*, 23 December 2012. http://www.guardian.co.uk/business/2012/dec/23/banking-year-of-shame-libor?intcmp=239

6 'Peer-to-peer lending boom could make banks obsolete', 17 December 2012, http://www.independent.co.uk/news/business/news/peertopeer-lending-boom-could-make-banks-obsolete-8421241.html

7 'How your pension is being used in a $6 trillion climate gamble', Bill McKibben and Jeremy Leggett, *Guardian*, 19 April 2013. http://www.guardian.co.uk/environment/blog/2013/apr/19/pension-6-trillion-climate-gamble?CMP=twt_gu

8 YouTube film of the Carbon Tracker 2013 report, *Bloomberg* HQ, 18 April 2013. https://www.youtube.com/watch?v=gn7T_AvPToY. This diary extract is compiled not from notes I took at the time – I was too busy chairing the discussion – but from written summaries that Paul Spedding, Steve Oman and Howard Pearce kindly provided me with for use in my account. I also used clips in their interviews for the YouTube film.

18 What next? The anatomy of the biggest crash

1 Renewables 2013 – Global Status Report', REN 21 (Renewable Energy Policy Network for the 21st Century). http://www.ren21.net/REN21Activities/GlobalStatusReport.aspx Modern renewables include all members of the renewables family except traditional biomass.

2 For a description of these events see *Half Gone*, my 2005 book published by Portobello, pp. 168–78. http://www.jeremyleggett.net/books/

19 The power of context: energy and security

1 'Renewable energy vital to our nation's security, economy', Lt. General Richard C. Zilmer (Retd.), *Renewable Energy World*, 9 November 2012. http://www.renewableenergyworld.com/rea/news/article/2012/11/renewable-energy-vital-to-our-nations-security-economy

2 'Powering America's defense: energy and the risks to national security', Center for Naval Analysis Military Advisory Board, May 2009. http://www.cna.org/reports/energy

3 For a full discussion of this concept, see Malcolm Gladwell, *The Tipping Point: How Little Things Can Make a Big Difference* (Back Bay Books, 2002).

4 'Renewables 2013 – Global Status Report', op. cit.

5 'US military to build an army of renewable energy plants', *Business Green*, 6 August 2012. http://www.businessgreen.com/bg/news/2197196/us-military-to-build-an-army-of-renewable-energy-plants

6 'US Navy defends "great green fleet" from Republican attacks', *Guardian*, 20 July 2012. http://www.guardian.co.uk/environment/2012/jul/20/us-navy-great-green-fleet-republicans

7 'IKEA plans for renewable self-sufficiency by 2020', *Recharge*, 23 October 2012. http://www.rechargenews.com/business_area/finance/article325897.ece

8 'Physicist reviews nuclear meltdowns', *New York Times*, 11 April 2011. http://www.nytimes.com/2011/04/12/science/12nuclear.html?_r=1&src=tptw

9 Kombikraftwerk, meaning combined power plant: http://www.kombikraftwerk.de/index.php?id=27

10 Keith Barnham, of the Physics Department at Imperial College London, Kaspar Knorr, of the Fraunhofer Institute for Wind Energy and Energy System Technology in Kassel, Germany, and Massimo Mazzer of the CNR-IMEM, Parco Area delle Scienze 37/A, 43124 Parma, Italy. They have a paper in press in *Energy Policy*: 'Benefits of photovoltaic power in supplying national energy demand'.

11 Mark Jacobson and Mark De Lucchi, 'A plan to power 100 percent of the planet with renewables', *Scientific American*, 26 October 2009. http://www.scientificamerican.com/article.cfm?id=a-path-to-sustainable-energy-by-2030

12 Amory B. Lovins, E. Kyle Data, Odd-Even Bustnes, Jonathan G. Koomey and Nathan J. Glasgow, *Winning the Oil Endgame: Innovation for Profits, Jobs and Security* (Earthscan, 2004).

13 I describe this experience in *The Carbon War*, op. cit., pp. 322–3.

20 The choice of roads: people and systems

1 Echoing the recommendation of the neuroscientists I heard at the Skoll Forum in 2010, I commend the work of Dan Ariely, and the audio tapes of the Skoll Forum sessions of 2010 and 2011, should the reader to whom this is new wish to delve deeper.

2 See, for example, chapter 4 in Dan Ariely, *Predictably Irrational: The Hidden Forces that Shape Our Decisions* (HarperCollins, 2009).

3 Rory Carroll, 'The temples of doom', *Guardian*, 28 October 2008. http://www.guardian.co.uk/environment/2008/oct/28/climatechange-population

4 SolarAid and SunnyMoney on *Newsnight*, YouTube film posted 25 February 2013. http://www.youtube.com/watch?v=efWlKLWPZ24&feature=youtu.be

5 Jeremy Leggett, 'A climate for change', *Time* and *Fortune* magazines, 16 November 2007. http://www.jeremyleggett.net/2007/11/shell-pays-for-articles-in-time-and-fortune-critical-of-their-world-view/

6 Described in *Half Gone*, op. cit.

Index